earth day
network

THE OFFICIAL

earth day
guide to
planet
repair

Denis Hayes
Chairman and CEO, Earth Day

A Lark Production

ISLAND PRESS
Washington, D.C. • Covelo, California

Library of Congress Cataloging-in-Publication Data

Hayes, Denis, 1944-
  The official Earth Day guide to planet repair / by Denis Hayes.
    p. cm.
Includes bibliographical references and index.
  ISBN 1-55963-809-5 (pbk. : alk. paper)
  1. Environmental protection. 2. Earth Day. I. Title.
  GE180 .H395 2000
  363.7–dc21

                                    99-050921

Printed on recycled, acid-free paper ✪

Manufactured in the United States of America
10 9 8 7 6 5 4 3 2 1

*To Gail, thanks for thirty years of creativity,
insight, iconoclasm, fun, and love*

Eco-IQ Quiz

Can you answer these questions? If not, read on to learn everything you need to know about global warming so you can help repair the planet.

1. What is the difference between a carbon sink and a kitchen sink? (See page 21.)

2. Why do ice sheets worry scientists? Why don't ice shelves? (See pages 18-19.)

3. Will strong CAFE standards affect Starbucks? (See page 131.)

4. What is the greenhouse effect? (See page 10.)

5. What is the most energy-efficient mode of transportation? (See page 70.)

6. Which produces more greenhouse gas emissions—your car or your house? (See page 77.)

7. What kind of light bulb is the energy equivalent of an SUV? (See page 98.)

8. How many people live on the earth today? (See page 3.)

9. How much faster than the human population is the car population growing? (See page 50.)

10. What is the source of almost all energy on earth? (See page 37.)

11. Do coal-fired power plants produce more or less greenhouse gas than plants that run on oil or natural gas? (See page 30.)

12. Does the Energy Star have anything to do with astronomy? (See page 80.)

13. Does a city resident have as high a chance of being mugged as a suburbanite has of being hurt in an automobile accident? (See page 72.)

14. How much energy does it take to dry your clothes on a clothesline? (See page 108.)

15. Do you Kyoto? (See page 135.)

Contents

Limited Warranty 1

Global Warnings 9
The Gathering Storm 10
Extreme Weather 16
Baked Alaska 18
Breathing Less is Not an Option 21
A Feverish Forecast 23

Refueling: Clean Energy 25
The Need to Refuel 26
Nineteenth-Century Energy: Carbon Fuels 28
Nuclear Power Play 33
Rays of Hope: Solar Energy 36
Turning to the Wind 38
Earth Friendly Fuel Cells 41
Biofuels 45

Operating Basics: Clean Living 47
Clean Transportation
Start Your Engines: Choosing the Best Car 49
The Safety Question 55
Urban Assault Vehicles 57
Electric Cars 59
Hybrid Cars: Today's Best Choice 61
Down the Road: Tomorrow's Cars 64

Beyond the Lonesome Driver: Transportation Alternatives . . 68
Move Closer to Work 71
Work Where You Live 73
Trains and Airplanes 74
Clean Homes
Shelter From the Storm: Efficient Buildings 77
Heating and Cooling 83
Clean Energy Checklist 89
Hot Water 92
Smart Lights 94
Chill: Refrigerators 100
Ways to Cook a Planet: Stoves and Ovens 102
Clean Plates: Dishwashers 105
Clean Clothes: Laundry 106
Power Shift 108
The Three Green Rs 111

Troubleshooting: Clean Power 117
America, The Skeptical 118
Programs, Not Promises 123
Ban Coal 126
Here Comes the Sun 127
Make That a Grande: CAFE Standards 131
Strengthen Appliance Efficiency Standards 134
Kyoto, or Bust 135
Nine Ways to Kyoto 138

Extended Warranty: Resources and Activism . . 143
A Brief History of Earth Day 144
Boycott Global Climate Coalition Companies 147
Elect Green Candidates 152
Earth Day Sponsors 156
The Earth Day Clean Energy Agenda 162
Join Earth Day Network 164
Support the Earth Day Foundation 165
Bibliography 166
Index 170

earth day
guide to
planet
repair

Limited Warranty

The environmental movement has displayed remarkable strength and intelligence since the first Earth Day in 1970. It has battled heroically to safeguard the world's health, diversity, and beauty, and it has been astonishingly successful. However, as the Earth's odometer rolls over into a new century, the Earth is facing a new threat—global warming—that dwarfs earlier perils.

We fought in the early years to establish the Environmental Protection Agency and pass laws to protect the air, the water, and endangered species. By Earth Day 1990, our efforts had spread around the world and our participants had increased tenfold. That international groundswell contributed greatly to the success of the Earth Summit in Rio de Janeiro in 1992, and led to the creation of new environmental protection agencies in many other countries.

In the United States, environmental concerns now influence the nation's investments, its lifestyles, and its laws. The right to a safe, healthy environment—a concept that essentially did not exist before 1970—has become a core American value. Today it enjoys wider, deeper public support than some values enshrined in the Bill of Rights.

Three decades after the first Earth Day, the bald eagle is no longer endangered and the Great Lakes are returning to life. Air pollution has decreased by more than a third, even though we now are driving almost twice as many cars more than twice as many miles a year. The Cuyahoga River no longer catches on fire, and hundreds of streams, lakes, and bays are swimmable. Millions of people choose to recycle, conserve water and energy, eat lower on the food chain, and limit their family size for environmental reasons. Earth Day has become a movement, not just a once-a-year event.

Most of this change has been immensely popular. Maybe too popular; too much save-the-earth chatter can drown out truly urgent messages and leave people feeling confused. Although polls show strong support for the environment among all income groups, educational levels, geographic regions, and ethnic backgrounds, the same polls contain a contradiction. Although most people believe that the global environmental problems that face us are real, they view these problems as too huge, too complex, and too abstract for them to do anything about. This Earth Day Guide was designed to overcome that inertia.

2

Earth Day helps the average person, the nonscientist who cares about the future of the Earth, learn how to have an impact. Here are a few of the daunting challenges that we need to face up to, soon:

- Most of the world's great biological systems are in a state of collapse because we have logged, trawled, or cultivated them to maximize short-term production. Both plant and animal species are disappearing at the fastest rate in 65 million years.

- The world's *existing* human population—six billion— is already three times as great as the planet's long-term carrying capacity if all people seek a level of affluence comparable to that currently enjoyed in, say, Sweden.

- We have carved two giant holes in the ozone layer, vastly increasing the exposure of people, plants, and animals to damaging radiation from the sun. Although the industrial North, the source of most ozone-destroying chemicals, has made important progress on this issue, much of the rest of the planet is not yet on board.

- We have raised the temperature of the entire planet and set in motion a series of inexorable forces that will raise it a lot more before we can stop the climb. Even if we now act decisively, it will take many decades to undo the damage.

Call to Arms

Our species has always altered its immediate environment. Ancient farmers converted the Fertile Crescent—the fabled Babylon—into the desert wastes of Iraq. But we have never before had the capacity to change the entire planet. In the last half of the twentieth century, for the first time in history, *Homo sapiens* became a geophysical force.

"We are modifying physical, chemical, and biological systems in new ways, at faster rates, and over larger spatial scales than ever recorded on Earth," Dr. Jane Lubchenco, former president of the American Association for the Advancement of Sci-

3

ence, recently warned the nation's scientific elite. "Humans have unwittingly embarked upon a grand experiment with our planet."

For example, there is little serious debate in the responsible scientific community about the reality of global warming. Yet the existence of this scientific consensus behind global warming comes as a surprise to many Americans, even literate Americans who pay close attention to the news. Every time they read a news story about a new global warming study, they encounter a quotation from someone saying that global warming is "unproven." Few readers notice that virtualy all the skeptical quotations come from the same half-dozen right-wing zealots. Similar misinformation is spouted by a handful of fanatics who think the world has too few people, not too many, or that CFCs don't destroy stratospheric ozone (although those who explained the chemistry have already been awarded a Nobel Prize for their discovery).

Countering this brownlash are the most prestigious scientific bodies in the world. The U.S. National Academy of Sciences and the U.K. Royal Society issued a joint paper in 1992 that stated, "The future of our planet is in the balance. Sustainable development can be achieved, but only if irreversible degradation of the environment can be halted in time. The next 30 years may be crucial."

This is not fear mongering by overwrought extremists, but a carefully phrased warning from the world's finest scientists. These scholars are trying to call public attention to the fact that the world has entered a dangerous new era.

Avoiding irreversible planetary calamity is the primary moral obligation of our time. This profound mission is what makes the modern environmental movement more than "just one more special interest."

It's a Very Small World

For 30 years the environmental movement has done many things well. It has enlisted superb scientists, teachers, managers, and lobbyists. It has produced some of the most creative and successful litigators in the history of American jurisprudence. Our best universities now teach courses in conservation biology, environmental law, and environmental engineering.

But we are now facing an utterly different set of environmental challenges, rooted in our global interdependence. Back in 1970, New England *states* sought to control sulfur emissions from upwind power plants in the Midwest. Thirty years later, *nations* now have a legitimate interest in the environmental behavior of other sovereign states. Greenhouse gases emitted in Germany affect the climate in Africa. Ozone-depleting chemicals used to clean computer chips in India contribute to skin cancer rates in Australia.

Many environmental issues involve thresholds that act as one-way valves: once you pass the threshold, there is no going back. To take some obvious examples, once a species becomes extinct, it cannot be restored to life. Once a denuded hillside loses all its topsoil, it cannot be reforested. Once a barrel of oil is burned, it is gone forever.

Energy and Global Warming

Earth Day 2000 will focus unprecedented international attention on energy and climate change—a particularly clear example of a problem that involves thresholds and that requires international cooperation. No other issue intersects with a wider variety of environmental problems than what kind of energy we employ to power society, where we get it, and how efficiently we use it. The wasteful use of outdated energy sources is producing climate change, oil spills, strip mines, nuclear waste, plutonium proliferation, smog, sulfate particulates, acid rain, childhood asthma, and myriad other environmental ills. Energy also addresses issues of social equity, national security, balance of payments, political influence, and oligopoly control. You don't have to be an expert on each of these issues, as long you get the general drift. Energy is a core concern of the modern world, and it is often a scary business.

Global warming—the greatest environmental consequence of mankind's current fuel choices—itself will cause many other environmental impacts: coastlines and rich agricultural river deltas will flood; marine and terrestrial species that cannot adapt to a swift temperature shift will die; agricultural productivity in the world's grain belts will decline; new diseases will spread in epidemic proportions; rainfall patterns will change; and so forth.

Although these issues are of global importance, they can be addressed at every level, from the family kitchen to the United Nations. Energy and global warming are enormously complex, yet there is a simplicity that lies on the other side of the complexity. *A huge majority of social and economic interests will win as a result of a transition to "green" energy sources and uses, while a comparatively small number of (very powerful) economic interests will lose.* Whenever progress has been made on a global environmental problem, it has been due to a worldwide outpouring of public sentiment that overpowers those with a vested interest in the status quo.

Today, most Americans enjoy the fruits of a robust economy powered by cheap fossil fuel. It's the best possible time to launch a national initiative to put our enviable technological ingenuity to the test and shift to low-carbon energy. Every country has sun, wind, and biomass (organic material such as firewood, paper, or orange peels that can be converted to fuel). By developing these resources, nations can inoculate themselves against unpredictable political winds in the handful of countries where most of the world's oil and coal are located.

But what can you, a single individual, do? Far more than you might think. This book offers common-sense advice for a two-pronged approach to saving the Earth: (1) take control of your own life, and (2) pressure politicians to change how they run the world. On the following pages, you will find a strategy to reduce your own energy consumption by half, almost without noticing, and learn how to encourage and take advantage of greener energy sources.

Some of these suggestions are pretty simple; other proposals are more challenging to our ingrained ways of thinking. Some you can do immediately, while others involve a commitment over time and even generations.

This book emphasizes solutions. You can show your kids how their personal choices, your family's lifestyles, and your government's position on renewable energy can have a positive—or negative—effect on the world. It's a simple legacy you can pass on by simply setting a good example.

How's Your Energy Level?

Earth Day, as a movement, is pushing for energy efficiency and for clean energy sources as the alternatives to global warming from wastefully burning carbon-rich fuels. The time is long overdue to accelerate this transition. Twenty-seven years after the first oil embargo, fourteen years after Chernobyl, eight years after the UN Earth Summit in Rio, and three years after the Kyoto Climate Conference, we have made little progress.

Every year, the world produces more carbon dioxide than the year before. Global carbon emissions have increased fourfold in the last half century. Emissions could easily quadruple again in the next half century, unless humanity mobilizes to prevent it.

The case for energy efficiency has long been made, but in recent years it has fallen on unhearing ears. Despite huge improvements in the energy efficiency of lights, appliances, buildings, industrial processes, etc., *per capita energy consumption in the United States is now back within 2 percent of the peak in 1973, before the first oil embargo.*

The reasons for this are straightforward:

- We buy bigger cars. Minivans, sport utility vehicles, and pickup trucks now comprise 51 percent of all new vehicle sales.

- We buy bigger homes. Houses have ballooned from an average of 1,600 square feet in 1973 to 2,100 square feet today, even though the average household has shrunk from 3.6 to three people.

- We own more electrical gadgets. Energy use by computers, video recorders, dishwashers, and clothes dryers has been growing 5 percent per year since 1973.

Industry achieved enormous improvements in energy efficiency per unit of output from 1973 to 1986, but then plateaued. As industrial output grew over the next decade, energy use grew along with it. (Fortunately, over the last two years industrial energy efficiency has once again improved.)

Existing technology can reduce energy consumption in most countries—and most definitely in the United States—by a factor of three or more. Energy efficiency saves money, benefits the economy, creates new jobs, and improves human health.

About the Book

The first step in solving problems is identifying them. Toward that end, the book's next section, **Global Warnings**, outlines the effects of global warming, from more bad weather to increased disease, from rising oceans to declining agriculture.

Refueling: Clean Energy explores the pros and cons of the energy sources we can choose from, and explains why Earth Day votes for renewable energy as the Earth's best hope.

The core of the book, **Operating Basics: Clean Living**, outlines simple things you can do to reduce your own energy use; **Troubleshooting: Clean Power** then lays out some things you can do to affect the world's energy use. For people of all economic backgrounds, Earth Day aggressively promotes national, local, and individual energy choices that produce no net carbon dioxide, no radioactive waste, and no materials that can be made into nuclear weapons.

Finally, **Extended Warranty: Resources and Activism** is a highly selective resource section for readers who are inspired to do more to keep the Earth alive and well.

That Earth Day has survived as an annual international event is a heartening testament to the strength of a good idea. It is also evidence that substantial numbers of us can transcend our troublesome tribal reflexes and embrace the reality that we all live in a place known as Downstream. So, as you read this book, I invite you to become part of Earth Day—every day. 🌍

Global Warnings

How much hot water are we in? This section tells how warm the globe has gotten, how high the water is rising, how sick we humans might become, and how fast the Earth's plants and animals are vanishing. It's not a pretty picture, but if we don't focus on it, it will only get worse.

The Gathering Storm

Scientists have determined that the five warmest years since the fifteenth century were all in the 1990s.

You've heard about global warming for years, but you may have only a vague sense of what it is. Global warming is caused by the *greenhouse effect*. Sunlight streams through the atmosphere, strikes Earth's surface, turns to heat, and then is radiated back toward outer space. But some of this heat is trapped by warming gases in the atmosphere and reflected back to Earth. These warming gases, such as carbon dioxide (CO_2), allow sunlight to pass in, but like the glass in a greenhouse, they block heat from escaping. As we pump more and more CO_2 into the air, more heat is trapped.

Your grandparents breathed air that contained about 280 parts per million CO_2. Your own most recent breath contained about 370 parts per million. The concentration is increasing by about two parts per million each year, and the rate of increase is accelerating. Within a few decades, you will breathe air containing *twice as much* CO_2 as the air your grandparents breathed unless we radically change our use of carbon fuels such as coal and oil. *Our air already contains more CO_2 than at any time since long before the first humans evolved.*

Scientists and world leaders acknowledge that the increase in greenhouse gases is already raining on our parade in the form of weird weather. The American Geophysical Society states that no other known phenomenon could explain the rise in temperature we've already experienced.

Heat is a form of energy. So, as the temperature rises, more energy accumulates in the atmosphere to power the Earth's great climate engines. More water evaporates more rapidly, later to be returned to the Earth in torrents. Some of this atmospheric energy is channeled into violent storms. Floods, monsoons, droughts, and hurricanes all are setting new records. For example, 1998 was the twentieth year in a row with a global mean temperature above the long-term average, and it had many extreme temperature-related disasters:

- 2,500 people drowned and *56 million* were driven from their homes in the Yangtze River flood in China.

- The monsoon season left two-thirds of Bangladesh under water for more than a month and rendered *21 million* people homeless.

- Hurricane Mitch hit Honduras with 180-mile-per-hour winds, washed away an estimated 70 percent of all crops, left 11,000 dead and a third of the population homeless. A year later, Honduras faced a major health crisis because Mitch had destroyed every sewage-handling facility in the country.

- 45 countries experienced severe droughts, many accompanied by runaway fires. The prolonged drought left Russia with its lowest grain harvest in 40 years.

- Healthy rainforests don't burn, but 1998 saw serious fires throughout Southeast Asia and the Amazon. Fires in southern Mexico were so extensive that they led to air quality alerts throughout Texas. Fires caused Florida officials to evacuate an entire county.

- In India, 3,000 people died of heat stroke.

Clearly, 1998 weather provided a "signature that the global warming we expected is rearing its head," as Kevin Trenberth of the National Center for Atmospheric Research put it.

If you are concerned that one year isn't enough evidence, scientifically speaking, don't worry. The world's climate scientists came to their belief in global warming after many decades of gathering information:

- Since record-keeping began in 1866, the 14 warmest years have *all* occurred after 1980. The temperature in 1998 was the hottest ever recorded, and represented the largest annual increase ever recorded.

11

• The amount of sea ice in Arctic waters has been shrinking an average of 14,000 square miles each year since 1978. In an article published in the journal *Science* in December 1999, the researchers attribute the ice loss to human-caused climate change. They calculate that the chance that the past 20 years' melting is due to normal climate variation is just 2 percent. They examined Arctic ice data for 46 years, and concluded that the chance that the last 46 years' melting is due to normal variation is just one tenth of a percent.

• To survive, plant and animal species are migrating farther north and to higher altitudes, following the migration and expansion of warmer climes.

• Extreme weather events have become more common. Unprecedented hurricane damage, winter floods, and summer droughts are wreaking serious havoc. More such disasters are expected, according to a report released by the International Federation of Red Cross Societies. The number of people needing aid after weather-related disasters such as floods and hurricanes rose from fewer than half a million in 1992 to 5.5 million in 1998. Ever-increasing damage from hurricanes has caused several insurance companies to refuse to issue policies in the Caribbean.

• Antarctic ice shelves, those parts of ice sheets that float on coastal waters, are melting. Over the past 50 years, the Antarctic Peninsula has lost 4,968 square miles from its ice shelves, and the pace appears to be accelerating.

• Throughout the tropics and mid-latitudes, glaciers are rapidly shrinking. The world's nonpolar glaciers have lost nearly half their ice this century. For example, 100 of the 150 glaciers that were in Montana's Glacier National Park in 1850 have completely disappeared, and the remaining 50 will vanish within the next 30 years if temperatures keep rising.

• 1998 brought the most massive wave of death ever recorded in coral reefs. In some reefs more than 70 percent of the tiny animals were killed. Warmer oceans harm the corals by affecting their food supply and increasing their susceptibility to human pollution as well as to viruses, bacteria, and fungi.

The National Oceanographic and Atmospheric Administration (NOAA) announced in December 1999, that this will be the second-warmest year for the United States in the twentieth century and projected 1999 to be the fifth warmest year for the whole planet since 1880. NOAA also released an assessment of the century's weather concluding that most of the worst weather events of the twentieth century occurred in the last ten years. James Baker, NOAA's head, concluded that, "Although there are some important things we still don't understand about climate change, the new data, the modeling results, and what we know about how the system works is even stronger in pointing toward the fact that we are seeing global warming."

CO_2 and You

Atmospheric CO_2, the principal culprit behind this destruction, has *only* increased by about 30 percent in the last century. But the rate of increase is accelerating radically as a larger human population seeks higher levels of affluence in booming economies.

If current trends are not consciously changed, CO_2 levels will *double* from the fairly stable, preindustrial levels within the next few decades. Carbon dioxide levels could double yet again—a fourfold increase—within another few decades.

The resulting warmth would likely accelerate the melting of Arctic tundra that has been frozen for eons, potentially releasing huge quantities of methane from the mucky, rotting vegetation. Methane is an even more powerful greenhouse gas than CO_2. Increasing levels of methane in the atmosphere would lead to still higher temperatures, speeding the melting of reflective ice caps and glaciers and causing a higher fraction of incoming sunlight to be absorbed and retained as heat.

If the world stays on its current course, agricultural productivity will plummet, hurricanes will become vastly more frequent and pow-

13

erful, and plant and animal extinction rates will soar even above today's levels.

When things get scary enough, people and nations tend to change direction. When it became clear that DDT was endangering birds, we banned it. When it became clear that CFCs were causing a huge hole in the ozone layer, we banned them. When it became clear that lead in gasoline was causing brain damage in children, we banned its use as a fuel additive.

But carbon dioxide is part of life. Plants need carbon dioxide to grow. Every time you exhale, you breathe out carbon dioxide—so we *can't* ban CO_2. Our only hope is to ensure that emissions of CO_2 from human activities are limited to levels that can be absorbed by trees and plants through photosynthesis. To do this, the world's inhabitants will need to find politically acceptable ways to reduce today's emissions levels by *80 percent.*

Eighty percent sounds impossibly high, I know. Not only that, changing the planet's climate is like piloting an ocean liner: we have to start turning the wheel long before the boat can begin to respond. Atmospheric carbon dioxide and other greenhouse gases have a warming impact that grows over time; once they are in the air they can stay there for centuries. Because oil and coal are so tightly integrated into the world economy, and because the full impact of CO_2 on climate has such a long lag time, we absolutely must start cutting CO_2 emissions *today.*

About now, you may be tempted to drop this book like a hot potato and comfort yourself with couch-potato behavior. Reality has no laugh track. But please keep reading.

The Good News

The really good news is that we know how to lick this problem. Europe and Japan use about *half* as much energy per unit of gross domestic product (GDP) as the United States, yet the Europeans and Japanese are comfortable, secure, and productive. And even they waste at least half the energy they use. Americans can achieve a 60-percent reduction in CO_2 by adopting a European standard of living while converting to the most efficient technologies currently being sold. By switching to solar energy, wind power, biofuels, and other renewable sources for more and more of our energy, we can indeed

14

CO$_2$ Chart

Here is a breakdown of carbon dioxide emissions from different fuel sources.

Fuel	Pounds of CO$_2$ per Million BTUs	Pounds of CO$_2$ per Standard Unit of Fuel
Electricity from Coal*	694	2.37 lbs CO$_2$/kWH
Electricity from Oil	628	2.14 lbs CO$_2$/kWH
Electricity from Gas	388	1.32 lbs CO$_2$/kWH
Wood**	216	2.59 tons of CO$_2$/cord
Coal (Direct Combustion)	210	2.48 tons of CO$_2$/ton
Fuel Oil	190	26.4 lbs CO$_2$/gallon
Gasoline	190	23.8 lbs CO$_2$/gallon
Natural Gas	118	12.1 lbs CO$_2$/therm

*Electricity from solar, wind, biofuels, and hydropower produces no CO$_2$.

**If wood is being used on a sustainable basis, growing new wood at the same rate as it is being burned, there is zero net CO$_2$ emission. Trees are removing as much CO$_2$ from the air as the burning wood is releasing.

achieve an 80-percent reduction in CO_2 emissions in our lifetimes.

America blazed the trail into this global warming dilemma. After Colonel E.L. Drake drilled the world's first oil well in Titusville, Pennsylvania, in 1868, we led the world into the oil age. Today, the United States, with only 4 percent of the world's population, emits 25 percent of the CO_2 produced by the entire world. It is high time for us to use our Yankee ingenuity and "can-do" attitude to steer our ocean liner of a country in a different, and better, direction. 🌍

Extreme Weather

The "control of nature" is a phrase conceived in arrogance, born of the Neanderthal age of biology and the convenience of man.—Rachel Carson

Climate means the generally prevailing weather in an area or region, averaged over a series of years. *Weather* is a volatile day-to-day phenomenon that changes much more rapidly—it is the wind, rain, temperature, pressure, and the like sweeping around you right now.

As the climate has gradually warmed, it has changed the world's weather in unpleasant ways. Higher temperatures cause more water to evaporate from the oceans, leading to more rain in flood-prone areas. Winters are warmer, so less snow accumulates. Instead of water trickling into rivers over several months as snow slowly melts, it flows directly down the mountains. Rivers become more seasonal. Hundred-year floods become annual occurrences. Higher temperatures also give rise to nasty events like twisters, thunderstorms, and hurricanes.

In the first eight years of the 1990s, insurance companies paid out a whopping $91.8 billion in weather-related claims—four times as much as they paid in the previous decade. The insurance industry blames its losses on both global warming and the increased number of people living in coastal areas and on flood plains. Extreme weather events around the world in 1998 forced more than 300 million people from their homes.

Our National Oceanographic and Atmospheric Administration (NOAA) says it is only a matter of time before the U.S. experiences a

$50 billion-dollar storm. That could force many insurance companies into bankruptcy.

It's easy to see why the insurance industry was the first big business to take climate change seriously and to lobby to stop it. In a greenhouse world, filled with hurricanes, floods, and droughts, insurance could become an endangered industry. Florida still hasn't finished mopping up after Hurricane Andrew (1993). So many insurance companies were washed away by Andrew that Florida taxpayers are now being forced to underwrite insurance for policyholders whom insurers would otherwise not touch.

A *reinsurance* company is one that contracts with many insurance firms to relieve them of part of their risk. In late 1998, Munich Re, one of the world's biggest reinsurance companies, predicted that large parts of the world, including the southeastern U.S., might become uninsurable.

When producing computerized predictions about the impacts of global warming, scientists have to make some assumptions about how much CO_2 humans will emit before we come to our senses. Most have examined what will happen if CO_2 doubles—but that may be the wrong assumption. Unless we swiftly undertake several bold initiatives, global CO_2 levels are likely to triple or even quadruple those from preindustrial times. There is no reason to believe that our children will all choose to limit their family size and dramatically reduce their CO_2 production. Unless a cadre of remarkable national leaders emerges around the world, leaders capable of building wide public support for a major change in global direction, a mere doubling of CO_2 is probably now unavoidable.

The Geophysical Fluid Dynamics Laboratory at Princeton University recently modeled, as best it could, a scenario in which CO_2 quadruples. It concluded, among other things, that soil moisture could drop by 50

> **T O O L B O X**
>
> *If you live in an area vulnerable to natural disasters, visit the National Renewable Energy Laboratory's Web site at* **www.nrel.gov** *and click on "Surviving Disaster with Renewable Energy."*

17

percent or more, wiping out much of agriculture in the Northern Hemisphere. The forces of nature that we have sought to master—wild rivers, tropical storms, gale-force winds, extremes of heat—would all increase in intensity. Our flimsy little dikes, dams, and storm shelters won't be much help.

Everyone talks about the weather, but unless you're a hurricane chaser or an extreme surfer, the forecasts are not encouraging. 🌍

Baked Alaska

Humans like to live near the coast; more than a third of us live within a hundred miles of a coastline. Storm surges already put about 46 million people at risk.

If we don't dramatically reduce our emissions of CO_2 and other warming gases, thousands of square miles of Florida could be underwater within this century. During the twentieth century, the mean global sea level rose by four to ten inches. By 2100, it could rise as much as three feet or more.

Where will all that water come from? To begin with, warm water expands to take up more space than cold water, so the same mass of water will fill more volume in a warmer world. The oceans are so deep that just a small percentage expansion provides a large additional volume.

More water will also come from melting alpine glaciers, which have already lost half their ice. But the largest sources of added water volume are the polar glaciers—particularly those in Antarctica. Ninety percent of the Earth's ice can be found in Antarctica, as well as 70 percent of all fresh water. Put this ice in perspective: if it were all to melt—which it won't—it would raise the levels of all the world's oceans some 200 hundred feet.

Antarctica's ice can be divided into three regions:

1. The East Antarctic Ice Sheet is the largest mass of ice in the world, and it has long been thought to be the most stable. Scientists therefore were disturbed to learn in late 1999, from radar images generated by a Canadian satellite, that a large number of unsuspected subterranean ice streams flow underneath the East Antarctic Ice Sheet. These can travel at speeds as fast as 3,000 feet per year and for distances of up to 500 miles. So East Antarctica, while far more stable than West Antarctica, is not as stable as everyone had assumed.

2. The Antarctic Peninsula is where many ice shelves are already disintegrating, apparently in step with global warming. Icebergs the size of Rhode Island are breaking free just offshore from the Peninsula. However, this is not affecting ocean levels because ice shelves already float on top of the water. Like ice cubes floating in a full glass of water, they do not displace more water as they melt.

3. The West Antarctic Ice Sheet is what keeps some scientists awake at night. That's because it is much less stable than the East Antarctic Ice Sheet. An ice sheet, unlike an ice shelf, rests on land. If it slips down and is launched (like a boat down a boat ramp) into the water, it raises the water level. While the West Antarctic Ice Sheet is far smaller than the East Antarctic Ice Sheet, it is still as big as Brazil and up to 7,000 feet thick. If the West Antarctic Ice Sheet should launch itself into the ocean, sea levels would rise as much as 20 feet all over the world.

New evidence suggests such a collapse happened at least once in the last 1.3 million years, when global temperatures were not much higher than they are now. The loss of the West Antarctic Ice Sheet would be an unparalleled human-caused disaster. It would flood vast tracts of low-lying land around the globe, displace tens of millions of people, and destroy the rich, rice-producing river deltas of East Asia.

Here are a few specific predictions of what might happen if we lose the West Antarctic Ice Sheet:

• San Francisco Bay and the adjacent Sacramento River Delta will double in surface area. Just a one-meter sea level rise would lead to property losses of $48 billion.

• If the high-level projections come true, most of New York City will be flooded (including all three major airports and the entire subway system). So will thousands of square miles of coastal Louisiana, including New Orleans.

• A one-meter sea rise would threaten the Japanese cities of Tokyo, Osaka, and Nagoya, where about 50 percent of that country's manufacturing is located. It would similarly threaten much of the economic base of Germany, the Netherlands, and Russia.

• Some of Bangladesh, the Netherlands, and many small Pacific island nations will *disappear completely*. Indonesia, the Philippines, and Malaysia will suffer major losses, as will agriculturally vital river deltas in China, Myanmar, Thailand, Vietnam, Egypt, and Nigeria.

When a tragedy's dimensions are too great for us to comprehend, we turn to humor. The threat of rising seas has led to lots of jokes about people in Nevada finding themselves with oceanfront property.

The reality, however, is no laughing matter. We must do whatever it takes to save the West Antarctic Ice Sheet—and with it many of our most-treasured places on Earth. 🌎

Breathing Less Is Not an Option

Woodman, spare that tree
Touch not a single bough!
In youth it sheltered me
And I'll protect it now.
—G.P. Morris, "The Oak"

Forests rank among our most-treasured places. They are the only home for many kinds of plants and animals. It is no surprise, then, that tree-planting campaigns are the most widely-touted solution for global warming. However, even the most ambitious global tree planting cannot begin to keep up with global CO_2 emissions.

Depending on how they are managed, forests can either pull carbon from the air or add more carbon to the atmosphere. Well-managed, they are probably the most reliable method we have of removing substantial amounts of carbon from the air and storing it for several decades.

Trees use solar energy to split carbon from carbon dioxide and hydrogen from water. Then they combine the carbon and hydrogen into solid materials, like cellulose, that make up the body of the tree. Roughly half the dry weight of wood is carbon. Trees live longer than most plants and therefore tie up CO_2 for longer periods. Forest litter can build up in the soil and act as a long-term carbon *sink*, therefore containing the carbon for even longer. Unfortunately, when trees are cut or burned more rapidly than they are planted, they have a negative effect on the build-up of atmospheric carbon.

Humans have already destroyed nearly two-thirds of our planet's original forest cover. Burning forests and the decomposition of forest vegetation put about 2 billion tons of carbon into the air each year (compared to about 7 billion tons from fossil fuels, such as oil and coal).

Tropical Rainforests

Most of the world's net forest loss since 1950 has taken place in the

21

tropics. Because of wildly irresponsible behavior by international timber companies, compounded by the massive impact of slash-and-burn agriculture, tropical rainforests are disappearing at a staggering rate.

Partly, this reflects the fact that many European and North American forests were cut centuries ago. But the continuing decline is also due to the fact that few cut or burned forests in the tropics are ever replanted. Of the estimated 14 million species on Earth, at least half live in tropical rainforests. Destroying these rainforests before we've even begun to understand their secrets is the biological equivalent of sacking and burning the Great Library of Alexandria.

A burning forest releases enormous amounts of carbon. Largely because of such fires, Brazil ranked fourth in CO_2 emissions in the late 1980s, behind only the United States, the Soviet Union, and China. Fires in the Amazon produced about 7 percent as much CO_2 as all the fossil fuel consumption in the entire world. In 1999, once again, several thousand huge out-of-control fires burned in the Amazon, completely overwhelming Brazilian officials in their efforts to contain them.

The Northern Forests

Most public attention has been focused on tropical rainforests, in part due to the efforts of Sting, Paul McCartney, and other celebrities. But focusing exclusively on the tropics can be misleading in terms of global warming. The planet's main land-based carbon sink is not in tropical forests—it is in forests that cover northern lands, particularly those of North America and Siberia.

The Siberian forest is the world's biggest. It has received far less attention than the Amazon, but it contains far more carbon—and it's being liquidated at an estimated rate of 10 million acres per year. In the wild confusion that accompanied the breakup of the Soviet Union, foreign timber companies from Japan, Korea, and the United States have been drawn to Siberia like buzzards to roadkill.

The temperate rainforests of America's Pacific Northwest are the most productive forests on Earth, and the coastal redwoods of California are the world champs at taking carbon from the air and storing it for long periods. The Pacific Forest Trust estimates that, with proper management (such as longer rotation times, more selective harvesting, and preserving all old-growth forests intact) the massive forest belt that stretches from Northern California through Southern Alaska might be able to store an additional 60 million tons of carbon per year. This, however, would require a major change in the policies of the region's timber companies, which currently measure the value of forests only in terms of millions of board-feet of harvested timber. Moreover, even if successful, this 60 million tons of sequestered carbon would equal only one percent of the *6 billion* tons emitted each year from the burning of carbon fuels.

The Sound of One Tree Falling

Let me repeat that: we dump *6 billion* tons of carbon into the atmosphere every year by burning coal, oil, and gas. There is no way we can plant nearly enough trees to soak it back up. With human populations soaring, it is not plausible that the world will turn its cropland back into nature preserves, or halt the slash-and-burn destruction.

Clearly, it's vital that we manage our forests and parks more intelligently. Forests provide habitat for diverse species, protect watersheds, retard erosion, and help absorb some carbon. Planting trees in our yards, along our streets, and in other vacant areas is also helpful. But we must never, not for a moment, be deluded into thinking that this is a meaningful cure for global warming. 🌏

A Feverish Forecast

If we go on the way we have, the fault is our greed and if we are not willing to change, we will disappear from the face of the globe, to be replaced by the insect.
—Jacques Cousteau

According to the World Health Organization, global warming is one of the largest public health challenges for the twenty-first century. Heat stress, the disruption of ecosystems, the spread of infectious diseases, forced migrations, loss of freshwater supplies, storm-related flooding, increased air pollution, and the disruption of agriculture will all have vast effects on human health.

Extreme heat, it seems, has a worse effect on health than other types of bad weather. When temperatures soar, mortality rates also soar. The elderly, the very young, and people who cannot afford air conditioning are particularly vulnerable. When heat is combined with high humidity, and when there is little relief after dark, the result is even worse.

Warmer, wetter weather will probably cause an increase in the spread of infectious diseases. Dr. Paul Epstein, a researcher and clinician at the Harvard School of Public Health, says, "Climate change is already a factor in terms of the distributions of malaria, dengue fever, and cholera." Disease-carrying mosquitoes, such as those that brought the West Nile-like encephalitis outbreak to New York and Connecticut in the summer of 1999, will spread further north because cold winters are often the factor that limits their territory; warmer winters mean a wider circumference of affected areas.

High temperatures in 1998 caused drought and forest fires in Southeast Asia as well as an infestation of pests. Mosquitoes flourished, spreading dengue fever in Thailand, and infecting many Indonesian and New Guinea villages with malaria. Parasitic worms became a significant problem in the Philippines.

The spread of cholera is another impending danger. The bacteria that cause this disease hide out in tiny marine animals called copepods that eat algae. Warmer water leads to more algae blooms; more algae means more copepods. Diseases carried by water snails, tsetse flies, crustaceans, sandflies, and blackflies will all spread to larger areas as global warming increases the territory warm enough for their hosts to survive in.

Global warming and the holes in the ozone layer are two distinct problems, but they feed on each other in complex ways. Global warming, for example, slows the rate at which the ozone holes close, exposing the planet to more decades of ultraviolet radiation. This radiation is a proven cause of skin cancer, cataracts, and sluggish immune systems.

The health effects of global warming will vary from region to region, and the science of predicting such regional effects is inexact. There is no dispute, however, that global warming poses serious threats to human health. Oil and coal companies are on notice that—like the cigarette industry before them—they may eventually be found liable for some of the illness and death they cause. The smartest of them are already worried. 🌏

Refueling: Clean Energy

Carbon fuels include coal, oil, and natural gas. They have powered everything from the industrial revolution to the information age. But even as we reduce some forms of pollution from individual cars, homes, power plants, and factories, our sum-total use of carbon fuels continues to grow—and to produce ever-increasing quantities of carbon dioxide.

The long-term solution is to move on from fossil fuels to smarter alternatives, just as we advanced from burning wood to burning oil. Many energy technologies now available were impractical, unavailable, or unaffordable just ten years ago. Fuel cells, for example, process hydrogen to obtain electricity, and produce only pure water as a by-product. Not long ago, fuel cells were thought to be too expensive for anything but space missions. Today, every major automobile manufacturer in the world is developing a fuel-cell vehicle, and hydrogen is emerging as the consensus candidate for the portable fuel of the future.

Solar cells, though deprived of the public sector support needed for a market transformation, have achieved dramatic price reductions and are now poised for take-off in the consumer marketplace. Wind is the fastest growing new source of electricity in the world. Biomass—organic matter that can be converted to fuel—is being used with greater efficiency and sophistication.

This section marks a path into the twenty-first century that isn't dependent on nineteenth-century energy sources.

The Need to Refuel

If benign, efficient technologies were applied, the world could permanently support a human population of two billion people in a lifestyle reflective of today's middle-class Europeans.—David Pimentel, Cornell University

Two billion people—70 percent of the population in the developing world—still rely on wood, animal dung, and kerosene for residential and commercial energy. These fuels, and a lack of access to electricity, have profoundly negative consequences for human health, economic development, and the environment:

- In these 400 million households, noxious fumes from interior fires are a serious health risk. Village huts often contain the most polluted air breathed by people any-where—sometimes ten times worse than the air in Mexico City or Beijing.

- Hundreds of thousands of lives are lost each year because rural health clinics are unable to refrigerate vac-cines.

- People without electricity are prevented from engaging in many income-earning activities.

- These and other consequences of a lack of electricity contribute to the great migration of the rural poor into third-world megacities, swelling already overcrowded slums.

Some governments have attempted to meet rural energy needs with centralized coal or oil-fired electrification programs. However, the huge up-front capital requirements of massive power plants, plus the daunting cost ($20,000 per mile) of running electrical wires to thousands of villages, make this impossible for most rural areas.

Even if such efforts were successful, they would in fact accel-erate global climate change and health-damaging pollution. According to the International Energy Agency, total energy use in the developing regions is forecast to surpass that of the major industrialized countries by 2010—with nearly all of it comprised of fossil fuels. Without a change in course, by then global carbon emissions could rise approximately 50 percent above 1990 levels.

Shortsighted, unwise choices are being made every day that lock segments of mankind into yesterday's energy sources for decades into the future—to our own detriment.

Renewable energy sources are *already* cheaper than conven-tional sources of electricity in most rural, third-world applications. *Decentralized* energy, which is generated right in the village where it is needed, completely eliminates the need for expensive transmission and distribution systems.

In a world of increasing economic polarization, sunlight is one of the most evenly distributed resources. It is in everyone's interest for the developing world to leapfrog over the carbon-fuel era directly into the solar-energy era.

But this won't just happen by itself. Consider the vast global enterprise already firmly in place that wants to sell more carbon fuels to developing nations. Western media broadcast an image of oil-fueled economic prosperity to the world. Without strong solar energy leadership by *and in* the United States, the developing world is likely to continue choosing nineteenth-century energy solutions to a twenty-first-century problem. 🌎

Nineteenth-Century Energy: Carbon Fuels

Mining is like a search-and-destroy mission.
—*Stewart L. Udall*

In 1997, the world consumed 5.2 billion tons of coal, 26.4 billion barrels of oil, and 81.7 trillion cubic feet of natural gas. Twenty private companies marketed more than a fifth of this fuel. If we include government-owned enterprises, the 20 largest companies marketed about half. Eighty percent came from 122 public and private companies scattered around the world.

The carbon-fuel-based energy system that powers the modern industrial world can neither be sustained nor replicated. As the world enters the twenty-first century, its richest countries cling to a set of energy sources appropriate to nineteenth-century levels of population and affluence. Coal—a filthy, bulky, dangerous-to-mine fuel that produces more greenhouse gas per unit than any other—should be outlawed as swiftly as possible. Oil, a wonderfully complex material, is too valuable to burn; it

28

should be increasingly saved for higher-value applications such as pharmaceuticals, lubricants, synthetic fibers, and plastics. And natural gas should be judiciously used as a bridge to a solar era.

Coal

The Earth contains a huge amount of coal, probably on the order of 7 to 10 trillion metric tons. Coal is our most abundant fuel. Enough coal lies in deposits that can be economically mined to meet the world's current level of energy use for hundreds of years.

Formidable environmental problems accompany both the extraction and the combustion of coal. Deep mines often acidify nearby streams, and they pose serious threats to the health and safety of miners. Surface mines, even when reclaimed, generally destroy the land for any agricultural use more complex than pastures. Coal combustion releases sulfur oxides and sulfates, mercury and other toxic metals, and carcinogenic organic compounds. Control technologies exist to reduce all these emissions dramatically, but hundreds of U.S. power plants have not installed them. If forced to comply with clean air standards, these power plants would be uneconomic to operate. In the developing world, even the most basic pollution controls are often absent.

Even otherwise sophisticated people confuse clean coal technologies with measures to solve global warming. A "clean" power plant, however burns just as much coal and emits just as much carbon dioxide as a dirty plant, while it sells less net power because the pollution control technologies

ENERGY EXTRAS

A draft report for the International Panel on Climate Change sketched out 40 different scenarios for carbon emissions in the year 2100—ranging from 4.3 billion tons to 36.7 billion tons—versus about 6 billion tons today.

The report makes clear that our choices of technologies will be as important as population growth and economic growth in determining the level of emissions.

used to scrub sulfur and toxic metals out of smokestacks gobble a great deal of energy. So the bottom line is this: *clean plants actually produce more global warming per kilowatt-hour than dirty plants.* (Dirty plants, on the other hand, produce more acid rain, more small particles that cause lung disease, and more poisonous mercury pollution.)

Coal combustion produces far more greenhouse gas than any other energy source. If we are serious about avoiding a world with higher seas, more hurricanes, increased disease, and diminished food, we must swiftly wean ourselves away from coal.

> ### ENERGY EXTRAS
>
> *The Chinese used coal for energy more than 2,000 years ago, as did the ancient Romans. Europeans used it in the brick, glass, and iron industries from the fourteenth century on. With the advent of the industrial revolution in the eighteenth century, coal use skyrocketed.*

Oil

In the mid-nineteenth century, there wasn't much of a market for oil. Hucksters sold it as a medical cure-all. It could also be burned in kerosene lamps, but most people preferred whale oil lamps. The United States then plunged into Civil War, cutting off the supply of whale oil. Kerosene lamps became more sophisticated, and the oil industry was born.

Sixty years later, in the 1920s, 80 percent of the world's energy still came from coal and a mere 16 percent came from oil and gas. Indeed, oil and gas combined did not pass coal as a global energy source until 1960—a century after the first oil well was drilled in Pennsylvania. Today, coal provides about 30 percent of energy from carbon fuels, oil contributes 42 percent, and natural gas about 28 percent.

Americans have some sense of the environmental problems associated with oil, especially in the wake of the 1989 Exxon Valdez oil spill. Some people may have even heard about the fights over oil drilling in ecologically sensitive areas in the Arctic and the tropics, as well as the thousands of toxic sites where service station gasoline tanks have leaked, Americans have some sense of the environmental problems associated with oil. As with many environmental problems, oil spillage is mostly a matter of scale. Hundreds of thousands of tons of oil are discharged into the environment every year from dispersed natural seeps and are easily digested by bacteria and other natural organisms. However, accidents and bad judgment (such as pouring used motor oil down storm sewers) are responsible for *millions of additional tons* of oil annually, often in ways that simply overwhelm the ability of nature to absorb the oil before it does serious damage.

ENERGY EXTRAS

The oil industry sputtered along for 50 years on sales for kerosene lamps. But with the advent of the gasoline-powered automobile, the industry boomed. The easily-tapped oil fields of Pennsylvania, Texas, Louisiana, and California became available just in time to power America's development into a modern industrial state.

Gas

Natural gas is much cleaner than oil, and vastly cleaner than coal. Burning it produces far less carbon dioxide than burning coal or oil. Natural gas is easily stored and transported, and it is the fuel of choice for most on-site generation of combined heat and electricity. It typically is the big winner when electricity is deregulated, and natural gas fired units will probably win more than 90 percent of the deregulated market, at least for a while. It also can be easily converted into methanol, which is a candidate fuel for the first generation of fuel-cell cars.

But natural gas is not without problems. It is often found in fragile areas where drilling for gas raises many of the same issues as drilling for oil. It is highly explosive, and pipelines must be built to exacting safety standards; human error leads to human tragedies. Liquefied natural gas vessels can explode like huge bombs—more

ENERGY EXTRAS

Humankind's energy options are often a result of conscious choice. In 1911, Winston Churchill decided to switch Britain's war fleet from coal to oil. This was a controversial and hard-fought decision. Britain was rich in coal and had no oil at the time. The coal industry, and the coal miners, were powerful forces. Experts considered the move unnecessarily risky and expensive. But Churchill believed that an oil-fired fleet would have the speed to defeat the German Navy at sea, and this strategic consideration trumped all other issues. Churchill's decision was a major turning point in the history of oil. Passenger ships and freighters swiftly followed the British Navy's lead, and within a few years, coal-fired vessels had largely disappeared.

powerful than any non-nuclear weapon in our arsenal—and they must be kept away from inhabited areas.

The most abundant source of carbon on the Earth is in a form of natural gas called methane hydrates. Methane hydrates are tiny, crystalline cages of ice containing methane molecules. If the ice melts, methane gas is released. Found over much of the deep ocean floor and also buried in the permafrost in Alaska and Siberia, methane hydrates contain twice as much carbon as all the coal, oil, and conventional natural gas resources on earth, combined. One of the great fears of global warming experts is that if sea level and Arctic temperatures rise sufficiently, the hydrates will melt and a huge amount of methane will escape to the atmosphere. This, in turn, will cause global temperatures to skyrocket still higher.

As a fuel to bridge the transition from carbon fuel to renewable energy, however, natural gas is, hands down, the most attractive option. And when the renewable-energy industry seeks influential allies in a world filled with rich, powerful enemies, it might look first toward the natural-gas industry. 🐾

Nuclear Power Play

...the maintenance of peace is a condition sine qua non
*for the widespread use of nuclear power which is
foreseen. A situation where power reactors above ground would
be the object of warfare from the air would
have unthinkable consequences, as would, for that matter,
fighting actions among some of the
100-odd warships propelled by nuclear power.*
—*Sigvard Eklund, 1973*
Director General, International Atomic Energy Agency

In 1974, the International Atomic Energy Agency (IAEA) predicted that nuclear power would provide 4.5 million megawatts of capacity by the year 2000. Nuclear power actually achieved less than 8 percent of that goal. Current installed capacity is 343,086 megawatts, and this is likely to be very close to an all-time peak. The U.S. Department of Energy now projects that world nuclear generation will decline by 50 percent over the next two decades. Of the 33 reactors now under construction around the world, about half will probably be cancelled for economic reasons before they are complete. The ailing industry is then likely to atrophy slowly as uneconomic and unsafe reactors are removed from service.

The demise of nuclear power has been caused by a spate of reactor accidents, the escalating costs of centralized facilities, and the declining costs of small, locally-situated alternatives such as natural gas combustion turbines. The last 20 U.S. reactors each cost $3,000 to $4,000 per kilowatt of capacity to build. New gas-fired plants cost $400 to $600 per kilowatt, and new wind turbines cost about $800 per kilowatt (and require no fuel). Other factors impeding nuclear power include worries about the long-term storage of radioactive waste, radiation risks, and the availability of uranium.

But the most compelling argument against nuclear power is this: any atom that can be split to provide commercial power can also be split in a bomb. Knowledge of how to build your basic

33

Nuclear power was the fastest-growing energy source in the world in the 1970s, increasing 700 percent in that decade. In the 1990s, nuclear power was the slowest-growing source, increasing less than 5 percent. In 1998, world nuclear-generating capacity actually declined by 175 megawatts. At the end of 1998, 429 nuclear reactors were operating worldwide, one fewer than five years earlier.

atom bomb, one capable of Hiroshima-level destruction, is now commonplace. Instructions can be downloaded from the Internet. The only missing element is a critical mass of bomb-grade uranium or plutonium.

To date, much of the spread of nuclear weapons, including the recent spate of explosions in India and Pakistan, has been made possible by commercial nuclear power programs. Indeed, securing fissionable isotopes for bombs was probably the ulterior motive behind the nuclear drive in several countries, such as Israel, that sought reactors ill-suited to their power needs.

Had the IAEA's earlier 4.5-million-megawatt prediction become reality, we would have had to build a *vastly* more stable political world, or we would have experienced global catastrophe.

The IAEA attempts to negotiate agreements with countries before allowing them to obtain nuclear materials. Even when it succeeds, such agreements can be ignored after coups, revolutions, or even elections. The last quarter century has seen many abrupt shifts of power within nations—shifts that can put nuclear weapons into the hands of virtually anyone.

In the early 1970s, Nobel Prize-winning physicist Hannes Alfven described the requirements of a stable nuclear state in striking terms:

Fission energy is safe only if a number of critical devices work as they should, if a number of people in key positions follow all of their instructions, if there is no sabotage, no hijacking of transports, if no reactor fuel processing plant or waste repository anywhere in the world is situated in a region of riots or guerilla activity, and no revolution or war—even a conventional one—takes place in these regions. The enormous quantities of extremely dangerous material must not get into the hands of ignorant people or desperados. No acts of God can be permitted.

Even world peace among sovereign states would not have offered sufficient stability in a world with 4.5 million megawatts of nuclear-generating capacity. The transport and storage of nuclear bomb-grade materials would necessarily have become a common item of commerce. Access by terrorists, and even by organized crime, would have been unpreventable. And this merely deals with intentional actions. Much of human history is the unintended consequence of incompetence. Nuclear technology leaves no room for mistakes.

As Earth Day focuses a global spotlight on the world's mounting consumption of fossil fuels, the nuclear industry will mount a slick, expensive campaign to argue that nuclear power is the answer. Indeed, many in the nuclear field believe that global warming will be the salvation of their dying industry.

When you come across the inevitable wave of pro-nuclear propaganda, remember that the most dangerous radioactive materials remain dangerous for eons. There is no natural cycle to reabsorb man-made nuclear materials like plutonium. Such substances can only be split into other, highly-radioactive elements with shorter life spans, or else safeguarded while they slowly decay. Plutonium has a half-life of 24,700 years. In other words, after about 25,000 years, half of the plutonium atoms in any piece of plutonium will have decayed into other elements. After another 25,000 years, half of the remaining plutonium atoms will have decayed (so one-quarter of the original amount will remain). After another 25,000 years, one-eighth will remain. After about 250,000 years, the original plutonium will be largely harmless. 🐾

Rays of Hope: Solar Energy

*The history of man is a graveyard of great cultures that came
to catastrophic ends because of their
incapacity for planned, rational, voluntary reaction
to challenge.—Erich Fromm*

As recently as 1980, solar energy was in the ascendancy. The solar energy movement had the missionary zeal of NASA in the early days, married to the brainpower of a Manhattan Project, but with the goal of producing BTUs, not bombs. Wind farms were sprouting all over California. Photovoltaics, a way of producing electricity directly from sunlight, was exciting Wall Street. Solar residential developments were springing up as new materials and designs made it possible to build attractive, affordable houses heated by the sun.

We threw it all away. Why? Because the Reagan Administration and its anti-environmental supporters in the oil and coal industries did not consider solar energy to be in their best interests. I used to joke that Interior Secretary James Watt and his henchmen suffered from deuteranopia, an eye disorder that prevents people from being able to see the color green.

The assault on solar energy began on the bleakest day of my professional life. That was the day I resigned as director of the federal Solar Energy Research Institute (now renamed the National Renewable Energy Laboratory). My resignation was a futile protest of the Reagan Administration's decision to slash my staff in half that afternoon, and to reduce the laboratory's annual budget by $100 million.

The ensuing blitzkrieg against renewable energy marked the beginning of an incredibly sad period for proponents of a clean energy future. In quick succession, federal and state renewable energy tax credits were abolished, as were many of the policies encouraging private research and development. In 1981 the United States unaccountably walked away from global leadership in the renewable-energy field.

For a brief period early in the Clinton Administration, it looked as though energy policy would again be steered back in a sensible direction. But the 1994 election brought in a new Congress suffering from group deuteranopia. Today, although the technology has advanced, the solar industry is arguably in worse shape than it was in 1980. And yet we as a nation continue to do nothing about it.

Shifting the world from carbon fuels to solar energy remains our best shot at holding the line against global climate change.

About 1.5 quadrillion megawatt-hours of solar energy arrive at the Earth's outer atmosphere each year. About a third of this is reflected back into space; about a fifth is absorbed by the atmosphere and drives the winds. Less than half reaches the Earth's surface. Still, this is about 10,000 times as much energy as humans currently use from all conventional sources combined.

No country uses as much energy as is contained in the sunlight that shines on its buildings. The sunshine that strikes American roads each year contains more energy than all the fossil fuels used by the entire world. Moreover, unlike fossil fuels, sunlight is a bottomless resource, not a limited stock. We can only burn a barrel of oil once, but the sun will be shining a billion years from now whether we harness solar energy or not.

Sunlight is ubiquitous and can be harvested in scattered, local facilities. Many solar applications can dispense with the expensive distribution wires needed by conventional power sources. Even as many developing countries are choosing to skip the era of telephone lines and proceed directly to cellular communications, they also could decide to skip American-style electrical transmission grids in favor of locally-harvested solar power.

> **ENERGY EXTRAS**
>
> *No nation includes the sun in its official energy budget. All other energy sources would be reduced to insignificance if it did. We think we heat our homes with gas, oil, or electricity. But without the sun, the air in those homes would be minus 240 degrees Celsius when we turned on the furnace.*

Some sunlight will be harnessed directly and used as electricity or stored as hydrogen. Solar energy also is captured by green plants through photosynthesis, and this stored energy can later be converted into solid, liquid, and gaseous fuels. Solar energy creates the winds—a formidable source of energy in some regions where they blow steadily. Solar energy, wind power, and green plants make attractive energy sources because most regions of the world have ample amounts of at least *one* of them. Deserts have little plant growth, for example, but intense solar energy; the far north tends to have little sun or vegetation, but it's very windy.

Our sun—a medium-sized star some 93 million miles away from Earth—burns 11 billion pounds of hydrogen every second in a massive thermonuclear reaction. It has been doing so for more than 4 billion years, and is expected to go on for another 5 billion years. Eventually, the sun will expand greatly, engulfing Mercury and Venus. Later, it will die down, contracting into a very small star (known as a white dwarf) no larger than Earth. Finally, it will burn out and become a black dwarf.

In other words, in the truly long-term future, we will need to think about another energy source, as well as another home planet. But for the next 5 billion years, the sun will do very nicely. 🌍

Turning to the Wind

The wind in North Dakota alone could produce a third of America's electricity.

Wind power is now the world's fastest-growing energy source. Wind is caused by the sun heating various parts of our spinning planet unevenly: land and water heat at different rates, as do mountains and deserts, the poles and the equator. Hot air rises, cool air rushes in, and wind blows. Wind patterns often are very predictable.

The sites for *wind farms*—vast arrays of windmills—must be selected carefully, after monitoring the wind for several seasons. Wind power increases with the cube of velocity. In other words, when the wind speed doubles, the power output increases eightfold.

The amount of electricity generated from the wind nearly tripled between 1995 and 1999. The 13,400 megawatts of installed capacity is enough to meet all the needs of about 5 million middle-class American households. Prices for installed turbines have fallen from $2,600 per kilowatt in 1981 to about $800 today.

America once led the world in wind energy technology, but we abandoned that lead during the Reagan Administration. Denmark has now seized global leadership, accounting for more than half of all new turbines installed everywhere in 1998. Danish wind turbine sales now equal the combined revenues of that nation's gas and fishing industries.

Wind with enough speed and regularity to generate reliable electricity is widely distributed around the world. India has more than 900 megawatts of turbines installed, and its ultimate wind capacity exceeds its total current power from all other sources. China has not yet focused on wind beyond a few demonstration turbines, but its wind-resource base is even greater. Just one stretch of largely vacant land in Inner Mongolia has enough wind to generate more electricity than China currently uses.

The American Midwest is richly endowed with wind resources—which is not a new discovery. Before rural electrification in the 1920s and 1930s, more than 8 million Midwestern windmills pumped water, made electricity, and ground grain. Carbon-fired power plants made these small windmills obsolete. But the new high-tech wind turbines that are springing up in fields from North Dakota to Texas will help make carbon-fuel plants obsolete.

Each of the new wind turbines can generate enough electricity for about 300 homes, and the fields around the machines can still be used to grow corn. In fact, wind could offer salvation for

---

ENERGY EXTRAS

*Scientists do not yet well understand the interactions
of birds and wind turbines. Some bird behavior
around turbines appears to humans
to be wildly irrational. It is indisputably
lethal. We should learn from our
experience with hydropower, which often has proven
deadly to fish runs, and minimize the harm
turbines do to birds. However, wild birds
and industrialized people have never gotten along
well. Birds are killed when they fly into cars,
trucks, power lines, office windows, and
airplanes. Some birds will certainly
tangle with wind turbines, to their detriment.
We must keep this number low but
understand that it will not be zero.*

---

hard-pressed farmers. An acre of windy prairie could produce between $4,000 and $10,000 worth of electricity per year—which is far more than the value of the land's crop of corn or wheat. And wind generation can be achieved without significantly diminishing crop production. In other words, the farmer reaps a double harvest. If we were to devote as much land to wind farms as to corn farms (3 percent), we could generate more electricity in the Midwest than America as a whole currently consumes.

While wind turbines and wheat cohabit comfortably, this is not always the case with birds. In mountainous regions, swift winds often attract raptors, some of which remain endangered species and all of which are magnificent. Bird deaths appear to be the largest environmental problem associated with wind development, and it is vitally important to choose sites that minimize avian mortality.

In the real world, wind power will never be our sole source of electricity, nor should it. Although the wind is always blowing somewhere, it is not always windy where power is needed. But wind can easily contribute at least twice as much electricity as the 9 percent Americans now get from damming rivers—and wind can do so while causing much less ecological damage. This is an industry poised for explosive growth. 🌿

Earth Friendly Fuel Cells

A gallon of gasoline weighs about six pounds. It is about 84 percent carbon, and the rest is mostly hydrogen. When burned, the carbon and hydrogen combine with oxygen to produce about 18 pounds of carbon dioxide and 8.3 pounds of water vapor.

Hydrogen (H_2) is the lightest and most abundant element in the universe. When hydrogen is burned or processed in a fuel cell, it combines with oxygen to release energy. Pure, drinkable water (H_2O) is the only by-product. Hydrogen is the cleanest, healthiest fuel possible.

Precisely because it is so light, hydrogen is never found in reservoirs as is oil or natural gas. Most of the hydrogen on Earth is already combined with some other substance, and energy must be spent to liberate it from the chemical bonds that link it to the other material. In fact, more energy is needed to split water molecules to yield hydrogen and oxygen than can be harnessed when the hydrogen and oxygen are later recombined. Hydrogen, then, is more a clean way to *store* energy than it is a natural energy *source*.

Seventy-five years ago, J. B. S. Haldane, the noted British scientist, gave a famous lecture at Cambridge University predicting that England would eventually turn to wind turbines and stored hydrogen for energy. Haldane's prophesy seemed almost whimsically foolish as the world turned more and more to oil, gas, and coal. But as we begin the new millennium, his 1923 lecture seems remarkably farsighted.

Most renewable energy sources are not continuous. Solar cells make electricity whenever the sun shines on them, but at night they produce no power. Wind turbines stop spinning when the wind stops blowing. Hydrogen can help fill in these gaps.

41

Most commercial hydrogen today is obtained by extracting it from fossil hydrocarbons. This produces carbon dioxide and warms the planet, thus undercutting the principal advantage of hydrogen. By obtaining hydrogen from water, however, a clean energy cycle can be developed that doesn't add to global warming. When the sun is shining, energy from the sun can split water to produce hydrogen and oxygen; the hydrogen and oxygen can later be recombined to run cars, generate electricity, or run furnaces in the middle of the night.

In addition to using solar electricity to split water, a variety of advanced technologies show promise of doing it more cheaply. These include photoelectrochemical and biological systems that produce hydrogen directly when sunlight falls on them. These technologies can work on both small and large scales. Someday you may be able to refuel your hydrogen car at home as well as at service stations.

Hydrogen can be stored as a pressurized gas, as a supercooled liquid, or in metal hydrides for use in vehicles. It can also be transported long distances in pipelines. Hydrogen pipelines, while not risk-free, are safer than natural gas pipelines.

ENERGY EXTRAS

The first generation of fuel cells now appears likely to be powered by methanol instead of pure hydrogen. If that methanol is produced from fossil fuels, it will add to global warming. If the methanol is produced from biomass, however, it will not add any net carbon dioxide to the atmosphere. Methanol (colloquially called "wood alcohol") is somewhat more expensive when produced from biomass than when it is made from coal or natural gas (since the costs of warming the planet are ignored). But the market for "green" methanol might soon be at least as large as the market for green electricity.

Fuel Cells

In the last few years, extraordinary progress has been made in the development of small, efficient fuel cells. Fuel cells combine hydrogen with oxygen, using catalysts, to produce an electric current and form water in the process. They can, quite literally, take us beyond the age of fire.

Although fuel cells have a long history, until recently they were too bulky and expensive to see widespread use. Their primary applications have been in space exploration (where cost is less important than reliability and efficiency) and in buildings (where size and weight are not as great a constraint as in cars). Their first applications in vehicles were in buses and submarines.

A Matter of Degrees

There are now five kinds of fuel cells, but three of them—alkali cells, molten carbonate cells, and solid-oxide cells—operate at such very high temperatures (up to 1,000°C) and face other such problems that they are not expected to see commercialization any time soon. The most promising early products are the most environmentally attractive, a rarity in the history of industrial innovation. The proton-exchange membrane (PEM) cell is most promising for automobiles, and the phosphoric acid cell is targeting the decentralized electrical generating market. Both are much cleaner than the others, which need very high temperatures. Although the higher-temperature cells are more efficient, even the low-temperature cells are 50 percent more efficient than today's internal combustion engines.

Both the cost and the size of fuel cells have shrunk dramatically. A fuel cell that required $33,000 of platinum for its catalysts in 1984 needs less than $500 worth today. In the last ten years, Ballard Power Systems has shrunk the size of its PEM fuel cell—ideal for automobiles—tenfold. Using the Ballard technology, DaimlerChrysler plans to sell 30,000 fuel-cell vehicles by 2004, and 100,000 by 2005.

Phosphoric acid cells are becoming more and more popular at places that need reliable power. By having units right on-site, hospitals, pharmaceutical plants, banks, and military installations, for example, can guarantee that their power will not go down, regardless of what happens to the electric grid.

With the deregulation of the electricity market, and increasing opportunities for private developers to sell their surplus power to the grid, various forms of on-site electrical generation are already becoming relatively commonplace. Today, small natural gas powered turbines (microturbines) dominate such do-it-yourself technology, but the fuel-cell industry has set its sights on this market.

> ENERGY EXTRAS
>
> *A fuel-cell car has all the advantages of an electric car—including excellent acceleration—without the major disadvantage of limited range. Instead of waiting hours to recharge a battery pack, a fuel-cell car can simply fill up with hydrogen and be on its way. This promise has prompted the world's major automobile companies to invest billions of dollars in fuel-cell development.*

One cloud on the hydrogen horizon is that the earliest, cost-effective fuel cells in both cars and buildings will not be run on hydrogen obtained from water. They will use hydrogen that has been stripped out of methane, methanol, or even gasoline. As a result, they will require devices that emit carbon dioxide. Until we harvest hydrogen from water (split by solar power) and make it widely available at service stations and in pipelines, fuel cells and fuel-cell automobiles will not reach their full potential.

However, the implementation of *any* fuel cells is a giant step in the right direction. Although the first generation will continue to produce some amount of CO_2, they will reinvigorate the dream of building a true, pollution-free hydrogen economy. 🌍

Biofuels

More than 40 percent of all waste in U.S. landfills is paper products. Another 7 percent is food and yard waste. Virtually all of this material could be converted into commercial fuels.

Biomass is any organic material that was recently alive. It includes such diverse stuff as firewood, paper, orange peels, and cow dung. Much biomass is burned directly by poor people around the world as a cooking fuel and to heat their dwellings. However, biomass can also be refined into biofuels and used for more sophisticated purposes. Biofuels can be liquid (like alcohol), gaseous (like methane), or solid (like charcoal). As global warming forces us to rethink our entire energy system, people are looking much more carefully at plans to create biorefineries that produce biofuels.

The most easily tapped biomass energy sources are the waste products of the food and forest industries. Bagasse, the residue from sugarcane, has long been used as an energy source in most cane-growing areas. The carbohydrate energy can be removed from corn and made into ethanol, leaving most of the protein intact to be fed to cattle. Municipal solid waste and gas from landfills are other sources of biomass energy.

The unharvested parts of food crops that are left on the field are a potential source of fuel, as long as the nutrients they contain are recycled to the land. So are animal wastes from feed lots. Combined with wind turbines, agricultural wastes could produce enough energy to make most farms self-sufficient.

Paper is another huge potential source of biofuels. Paper, of course, should be recycled several times before it's discarded. But when paper fibers get so ragged they can no longer be reused, they should be tapped for their energy content. Many paper mills once produced all their own electricity, and a surplus to boot; now, some are once again exploring the possibility.

Where does the energy in biomass come from? Green plants

capture sunlight through photosynthesis. In photosynthesis, plants combine carbon (taken from carbon dioxide in the air) with hydrogen (taken from water) to form new molecules that become the body of the plant. This new plant material, or biomass, stores the captured energy from the sun in chemical bonds.

During the course of hundreds of millions of years, some biomass (and the carbon it contained) was slowly converted into fossil fuel. When coal or oil are burned, carbon that may have been removed from the air millions of years ago is again returned to the atmosphere as carbon dioxide, and it contributes to global warming. This fossil carbon begins to re-create the atmosphere, and the climate, that prevailed when dinosaurs ruled the Earth.

As explained earlier, fossil carbon is being burned to produce carbon dioxide incomparably faster than the rate at which CO_2 was originally locked up in fossil fuels. Carbon that plants spent a million years removing from the air can be returned in a single year of fossil-fuel use. So it is far better for the planet to burn biomass than it is to burn fossil fuels. If we grow as much new biomass as we burn in biofuels, the carbon in the atmosphere will remain in equilibrium. It is only when we release carbon that was locked up fairly permanently in fossil fuels—or we harvest forests and grasslands without replanting them—that we alter the makeup of the atmosphere.

If we are to live in a diverse, healthy, resilient world, we cannot harness most of the Earth's plant growth to power human activities. Indeed, human demands on the world's last great forests and grasslands are already triggering an epidemic of extinction. Nevertheless, it makes great sense to tap some biomass for commercial energy—especially the biomass already contained in human waste streams. 🦖

Operating Basics: Clean Living

The Kyoto Treaty calls for the
United States to emit 7 percent less
global warming
gas in 2010 than we emitted in 1990.
The emissions of the United States
are the total emissions of our
vehicles, our buildings, our power plants,
and our factories.

The government could do many things to help us achieve that goal if our leaders had the guts to stand up to the oil and coal industry. *But there is nothing the government can do to stop us from beginning to meet this goal in our own households, cars, and businesses!*

Several institutions, like the Interface Corporation (a Fortune 500 carpet manufacturer), Patagonia (a manufacturer of athletic and mountaineering equipment), Stonyfield Yogurt, Tufts University, and the World Resources Institute have decided to implement the Kyoto Treaty without waiting for the government to act first.

You can too.

Personal choices are not a substitute for political action. In truth, there are no "50 Simple Things You Can Do to Save the Planet." Global problems can only be comprehensively solved through big changes in public policies that engage most people in the shared enterprise.

Similarly, however, political action is no substitute for leading lives that reflect environmental values. As long as people keep buying energy hogs like Sports Utility Vehicles, manufacturers will keep making them.

This section of the book explores common-sense things each one of us can do to reduce dramatically greenhouse emissions from our homes, cars, businesses, schools, and so forth. These suggestions also happen to save money, improve health, and reduce stress. That's a big bang for your environmental buck!

Clean Transportation

Start Your Engines: Choosing the Best Car

America has far more licensed drivers than registered voters.

A typical American car uses only about 2 percent of the potential energy in its gasoline to move its driver from point A to point B. That appalling fact should be a source of deep embarrassment to the automobile industry. Hobbyists and college students routinely build cars that meet all federal safety standards, have high performance characteristics, and get between 100 and 200 miles per gallon. Efficiency is not rocket science.

For some years, I have been waiting for a commercial car that:

• averages at least 55 miles per gallon (double the federal mileage standard);

• has excellent brakes, dual air bags, and other reasonable safety features;

• exhibits enough spunk to maneuver smartly in traffic;

• has a driving range of at least 300 miles between refills or recharges; and

• is well engineered and durable.

There are finally two cars on the road that meet my requirements. They are described later, in the section on hybrid cars. I realize that hybrid cars will not appeal to everyone, so let's first look at other options.

Where We Are Coming From

My great-grandparents never rode in an automobile. In 1900, few people had. But by mid-century, 50 million cars were on the road. Today, there are 500 million, worldwide, and more rolling out of the factories every hour. *At the dawn of the new millennium, the car population is increasing five times as fast as the human population.*

While it is true that our cars produce less pollution per gallon than they did 30 years ago, that progress is more than offset by the increased number of cars, the increased number of miles we drive today, and the congested stop-and-go conditions under which we drive them.

The baby-boomer generation will remember when we used to brag about how many miles per gallon our cars got. But a strong economy and cheap gasoline have erased gas mileage as a concern. These days, a bulging, overpowered sport utility vehicle, loaded with options, seems to be the vehicle of choice.

Earth Day's wish list includes permanent improvements in every car's fuel efficiency. By increasing awareness of the environmental costs of our driving habits, we aim to increase public demand for more fuel-efficient vehicles and send an unmistakable message to car manufacturers. We want to hold sport utility vehicles, pickup trucks, and minivans to the same mileage standards as regular cars. At the moment, manufacturers are enjoying their biggest profit margins on these monster vehicles.

America pledged at the Rio Earth Summit in 1992 to lower its CO_2 emissions to 1990 levels by now (the year 2000). Carbon dioxide emissions from cars arguably can be reduced more easily than emissions from any other source. The automobile lobby is very strong, however, and too many drivers express their indifference to global warming by buying sport utility vehicles (SUVs) with terrible mileage. The upshot? Carbon dioxide emissions from transportation increased by 10 percent between 1990 and 1997, and they continue to increase.

Global warming is not the only peril posed by our gasoline binge. Our oil addiction leads to other regrettable consequences, like the Desert Storm War. If Kuwait's largest export were broccoli, Saddam Hussein wouldn't have bothered, and

Americans would not be spending countless billions of tax dollars to keep a full-time fleet in the Persian Gulf.

Drilling for oil destroys delicate ecosystems, from the Arctic tundra to tropical rainforests. At the time of the Arab oil embargo in 1973–74, the United States imported 36 percent of its oil. Now it imports 56 percent. With huge volumes of petroleum being shipped around the world, pipeline leaks and tanker accidents are inevitable.

Our cars are stunningly inefficient. The average new car sold today in America has worse fuel efficiency than the older car it replaces. This is due in large part to the explosive growth in sport utility vehicles, vans, and light trucks. Americans buy only 2 million small cars a year, compared to 5 million a year in Europe (where gasoline costs $4 to $5 per gallon).

What Is Important to You?

We tend to buy cars for reasons of style or prestige, giving little thought to how their design features match our actual needs. Yet your car is probably the second-most-expensive thing you will ever buy. The average new car price equals the total earnings for 25 weeks of the median American family. Most people borrow to finance their cars, and the average auto loan is now for 56 months. That translates into 10 to 12 percent of family earnings for almost five years just to pay off the loan. On top of this, you must factor in the cost of repairs, maintenance, gasoline, insurance, license fees, parking, and tolls. Based on the numbers alone, clearly a car should never be an impulse purchase.

Think about what is really important to you. How often do you expect to hit 140 miles per hour in your muscle car? Do you plan to take your SUV on safari? Will your car be used mainly for long trips or short commuting hops? Do you frequently transport many passengers? How often do you haul freight? How much serious camping do you do? If you seldom need a large vehicle, wouldn't it be more sensible to rent one on an as-needed basis than to buy one and haul two tons of metal with you wherever you go? Unless you feel you have to impress clients or potential mates with a designer vehicle, try to stay focused on a car as being nothing more than a transportation appliance. Your car need say no more about your status or virility than your washing machine.

Once you have decided what kind of vehicle you need, pick one that is kind to the atmosphere. The best measure of how much your car heats the planet is how many miles per gallon (mpg) it gets. The higher the mpg, the less carbon dioxide your car produces per mile. The most fuel-efficient car in a size category typically uses much less gasoline than the least efficient one in the same size category.

The Environmental Defense Fund has created an Internet-based calculator that will allow you to figure out how much CO_2 your particular car puts into the atmosphere, whatever its make, model, and year. You can access it at *www.edf.org/cgi-bin/TailpipeTally.pl.*

The Best Cars

The most important environmental car buying advice is to buy the most efficient vehicle that meets your genuine needs, and drive it no more than necessary. A young or old couple living alone may enjoy a two-passenger Honda Insight. A larger family, perhaps with an elderly or disabled parent to care for, may absolutely need a van. In each size category, there are both better and worse vehicles to choose among. This list reflects a personal judgment about the inevitable trade-offs between fuel efficiency, safety, reliability, and performance.

Two-Passenger Car
Honda Insight*
Mazda Miata

Subcompact Car
Honda Civic GX or HX
Mitsubishi Mirage

Compact Car
Toyota Corolla
Saturn SL
Toyota Prius**

Small Wagon
Saturn SW
Suzuki Esteem

Midsize Car
Honda Accord
Toyota Camry

Midsize Wagon
Suburu Legacy Wagon
Volkswagen Passat Wagon

Large Car
Toyota Avalon
Dodge Intrepid

Minivan
Dodge Caravan

Large Van
Volkswagen Eurovan

Compact Pickup
Ford Ranger
Toyota Tacoma

Standard Pickup
Mazda B4000
Chevrolet C1500 Silverado***

Small Sport Utility Vehicle
Toyota RAV4
Suzuki Vitara

Larger Sport Utility Vehicle
Toyota 4Runner
Jeep Cherokee

*The two-passenger Honda Insight, which gets 61 miles per gallon in the city and 70 miles per gallon on the highway, is the most efficient commercial vehicle on the American road in 2000.

**Available in mid-2000.

***As distinct from the C2500 or the K1500 Silverados, which get terrible mileage.

Avoid These Cars

Whatever their other merits, these vehicles have such terrible mileage that they should be an acute embarrassment to the automotive engineering profession. There are comparable vehicles with much better fuel efficiency.

Vehicle	City Fuel Mileage	Highway Fuel Mileage
Ferrari 550 Marane	9	14
Rolls Royce Silver Spur	10	15
Bentley Brooklands/ Ford Excursion*	10	15
Land Rover Range Rover	12	15
Cadillac Escalade	12	16
Chevy Tahoe/ GMC Yukon K1500	12	16
Ford Excursion	12	16
Dodge Ram B1500	12	17
Lincoln Navigator	12	17
Toyota Land Cruiser	13	16
Lexus LX470	13	16
BMW X5	13	17

*Estimate. The EPA had not posted the official figures at the time this book went to press.

Before you buy your next car, check out the federal government's fuel efficiency Web site *http://www.fueleconomy.gov*. It is easy to navigate and has up-to-date information on all the vehicles the federal government has fuel economy information about. Also review *ACEEE's Green Book: The Environmental Guide to Cars and Trucks* for a more thorough environmental assessment of individual vehicles, as well as advice on driving and maintenance. The American Council for an Energy-Efficient Economy's Web site at *www.aceee.org/greenercars* also contains useful guidance.

Think about what your purchase means for the planet.

The Safety Question

Excellent brakes are your best single defense in a hostile, uncontrollable highway environment. Antilock brakes are always worth the extra cost.

Every few years, you read about a study that shows a statistical correlation between small cars and fatal accidents. The report always states that a high percentage of vehicle fatalities involves small cars. It then concludes that small cars are inherently unsafe, and that smart people should buy tanks.

That is an unhappy marriage of misleading statistics and dumb policy.

The statistics are misleading because they emphasize size (small) instead of cost (cheap). Let me explain. In America, small cars are cheap cars. As a result, they are more affordable to young people, who tend to drive more recklessly and to be less mature in their use of alcohol, and who consequently have more high-speed accidents than their elders do. Cheap cars also tend to be less sturdily constructed with fewer and poorer safety features than large, expensive cars. For example, small cars (being inexpensive) usually have inferior brakes.

No wonder studies "prove" that fuel-efficient cars are dangerous. *But there is no reason that a lightweight, efficient car cannot be a safe car.* A well-constructed small car can compensate for its lack of mass with intelligent design, state-of-the-art materials, electronic safety features, good tires, maneuverability, and antilock brakes. For example, adjustable foot pedals allow short drivers to distance themselves from the steering wheel, making it less likely for them to be injured by the steering column in the event of an accident. Before inflating, new "smart" air bags instantly calculate the seat position, the severity of the accident, and whether the seat belts are buckled.

Materials, structural design, and special features all are crucial when it comes to safety. For example, the new Cadillac DeVille features night vision, an extra that automobile manufacturers don't make available on small cars. The night vision device senses infrared heat signatures from animals and cars and projects them low on the windshield, below the driver's main line of vision. Night vision claims to increase the distance a driver can see ahead up to 500 percent. It is pricey—more than $1,000—but then so were antilock brakes and air bags when they were first introduced.

Race-car drivers in lightweight vehicles survive crashes at 200 miles per hour because racing cars absorb energy in crashes rather than transfer it to the driver. The carbon fiber used in these cars could also be used in passenger vehicles. Lightweight composite materials not only save energy, but also provide better protection than the steel they replace. Composites are strong and bouncy. Just ten pounds of hollow, crushable carbon-fiber-and-plastic cones can absorb all the crash energy of a 1,200-pound car hitting a wall at 50 miles per hour.

People who buy sport utility vehicles often think they are buying Volvos on steroids. Crash-test results show that this is not a safe assumption. Some SUVs do surprisingly poorly.

Size *does* matter, of course. In a collision between a 5,000-pound SUV and a comparably equipped 2,500-pound subcompact coupe going the same speed, simple physics tells you that the coupe will get the worst of it. But, as we have seen over the last ten years, buying additional tonnage for safety just leads to an automotive arms race. Smith owns an Escort. Then Jones buys a Taurus. So Smith buys an Explorer. Jones counters with an Expedition. Smith fires back with an Excursion. Pretty soon, everyone is driving Bradley Fighting Vehicles, and no one is safe.

A smart, defensive driver in a small car has a huge advantage over a dumb, sloppy driver in a Land Rover. If you stay alert—and sober—when you are behind the wheel, the odds will favor you.

Buy safe cars, but don't buy unnecessary tonnage. 🌍

Urban Assault Vehicles

*How serious is carbon pollution and global warming?
Let's just say that if China or India ever embraces America's
sport utility vehicle craze, we're all done for.*

Historically, Detroit priced cars so that the larger the car, the higher its price. It cost only a little more to build a Cadillac than a Chevrolet, but GM sold the Caddy at a much higher price. Size, prestige, and price became tangled together in buyers' minds.

In a recent stroke of marketing genius, auto manufacturers extended that bigger-is-better marketing mindset to the sport utility vehicle. They upgraded the interiors, installed automatic transmissions, made it possible to lock the hubs for four-wheel drive without stepping outside, installed top-of-the-line sound systems, and repackaged SUVs as boxcar-sized luxury vehicles.

Sport utility vehicles, once a rugged niche owned by unpadded Jeeps and Land Rovers, is now full of Mercedes, Lincolns, BMWs, Lexuses, Infinitis—and even Hummers. Their owners pilot these fuel guzzlers around suburban mall parking lots, not the Serengeti Plain. When an ordinary car gets into an accident with an SUV, it is like being hit by a cement truck. The new Ford Excursion actually employs the automotive equivalent of a railroad "cow catcher" by its front bumper, so that it doesn't ride over small, low-to-the-ground cars and crush them. Insurance agents have begun referring to these behemoths as Urban Assault Vehicles.

The Ford Explorer spawned the Ford Expedition, which in turn has given birth to the new 18-foot, 11-inch Ford Excursion. If the Excursion were an inch wider it would need running lights on top and down the side. It gets a miserable ten miles to the gallon in city driving. Its Valdez-sized fuel tank holds 44 gallons; in Europe it would cost up to $220 to fill 'er up. The Excursion is designed for nine people (at a time when the American family has shrunk to fewer than four). It weighs as much as three Honda Civics and has as much cargo room behind its third row of seats as the combined trunk space in three average cars. An

Excursion with a 5.4 liter, eight-cylinder engine, four-wheel drive, and an automatic transmission pumps out 20 tons of greenhouse gas every 15,000 miles.

To address the demand for increased comfort, the SUV market is moving toward crossover vehicles that are based on cars rather than trucks. For example, the Lexus RX300 sits on the chassis of a Toyota Camry sedan. The BMW X5 is based on the BMW station wagon. They are relatively worthless in off-road conditions. But industry estimates show that barely 10 percent of all SUVs ever venture off-road. The percentage is even lower in $60,000 SUVs with luxury nameplates.

People are surprised to learn that SUVs are designed principally to appeal to urban women. In the 1990s, women make most car-buying decisions, and are attracted to masculine models explicitly advertised as chiseled hardbody vehicles with triceps around each wheel. (Lest my fellow males start feeling a little superior, we will never live down the 1953 Cadillac with two over-sized front bumper guards, widely known as the "Dagmar" model after a blonde television starlet.)

By positioning the sports utility vehicle as a prestigious source of rugged safety in a hostile urban world, the auto industry has sold millions of SUVs to upper-middle-class women. And by filling SUV advertisements with pictures of the Rocky Mountains and various national parks, the industry has managed to make this environmental horror story the favored vehicle of the greenest cohort of American consumers—well-educated women.

Of course, there are legitimate purposes for SUVs. If your life routinely takes you into sand, mud, snow, or pastures, four-wheel drive is important. SUVs are perfect for big dogs; neither you nor your Great Dane will be happy sharing a Honda Civic. However, if you feel you really need an SUV, consider buying a used one—back from the days when they were genuine utility vehicles—modified trucks with four cylinders and decent mileage. A ten-year-old, four-cylinder Toyota 4 Runner that gets 30 miles per gallon on the highway probably still has half its useful life ahead of it. It can tide you over until the industry starts making efficient SUVs, perhaps powered by fuel cells. 🌍

Electric Cars

*If everyone bought the most efficient car in each
class, the U.S. would save 1.47 billion gallons of gasoline
every year.*

Six of the seven greenest cars in 1999 are electric cars. For people who make a lot of short trips and live in a warm or moderate climate where batteries perform best, an electric car is cool.

Most electric cars can go from 50 to 100 miles on a single charge of their batteries. The average U.S. car travels about 27 miles a day, so an electric car can be a clean, efficient choice, especially as a second vehicle used for urban commuting. Most of today's electric cars use regular lead-acid batteries, but some have more advanced batteries, e.g., nickel-metal hydride or lithium-ion. Even more advanced batteries with longer lives, faster recharging times, and greater energy densities are on the horizon.

Electric cars have been a cottage industry for a long time. Most manufacturers are tiny by the standards of the automobile industry. They have great difficulty raising capital, because investors are reluctant to invest in companies they see as competing with some of the largest corporations in the world such as Ford, Toyota, or DaimlerChrysler.

When General Motors entered the electric vehicle world in the mid-1990s with its EV1—carrying the implicit promise of an E(lectric) V(ehicle)2, an EV3, etc.—electric vehicle entrepreneurs were of two minds. The elephant was now in the living room, and they could no longer tell investors that they were aiming at a niche that GM would ignore. However, GM's entry was widely expected to validate the electric car category, much as IBM's personal computer made the desktop computer viable in corporate boardrooms and the mass market.

This hasn't happened, despite the fact that the EV1 is vastly better as a car than the first IBM PC was as a computer. The EV1 can go from zero to 60 in less than nine seconds. With a drag coefficient of 0.19, it is the most aerodynamic commercial vehicle on the road. The EV1 has cruise control, traction control, dual airbags, power windows, and a heat-pump climate-control system. It has antilock, regenerative brakes that produce electricity to recharge the battery every time you step on them. It has a lightweight aluminum alloy structure and impact-absorbing composite exterior panels. With a nickel-metal hydride battery pack, the EV1 gets 75 to 130 miles per recharge in the real world (depending on terrain, driving habits, temperature, etc.).

The EV1 is a wonderful urban short-trip vehicle. But GM initially limited sales of EV1s to Los Angeles—a city famous for its long distances between travel destinations. Over the years, GM began selling the vehicle in just five other cities in California and Arizona. It has sold only a few hundred EV1s.

After driving an EV1 in 1996, I learned that GM had no plans to sell the vehicle in Seattle, where I live. I asked some top-level GM officials whether they would make the car available in Seattle if I personally rounded up a 100 qualified buyers. With some awkwardness, the answer was "no." No one has ever been able to give me a reason. Seattle—with its hills, congestion, mild weather, green consumers, short commutes, and dirt-cheap electricity—is a

---

T O O L B O X

An excellent site to learn about electric cars is http://www.evaa.org, *the Electric Vehicle Association of the Americas. If you get serious about buying an electric vehicle, try* www.ElectricCars.com. *It is a membership site costing $19.95 for 36 months in the U.S. ($29.95 Canadian), so its access is not slowed down by lots of people who are not really interested. It contains a wealth of information on virtually every electric vehicle that exists and sends out regular e-mails to members about new developments.*

nearly perfect city for electric vehicles. But GM's marketing geniuses placed all their bets on California and produced a billion-dollar belly flop.

Its electric foray cost General Motors more than one billion dollars. Sadly, GM appears to be learning the wrong lesson. Most GM officials seem to believe that the EV1 was simply priced too high for a two-passenger car. But most two-passenger sports cars cost more than the $30,000 EV1— some a lot more. The real lesson is that GM should have used a more advanced battery system with a longer range, and it should have targeted its real market much more intelligently. Limiting electric vehicle sales to Los Angeles virtually guaranteed failure.

The EV1's failure was so total, and so expensive, that it has most likely wrecked the entire market for electric vehicles by major automobile manufacturers. The excitement today in GM's new headquarters in Detroit's Renaissance Center is all with hybrids, fuel cells, and GM's new technology-sharing agreement with Toyota. Many of the technical innovations of the EV1 will survive, but probably in a reconfigured vehicle with a hybrid engine. 🐾

Hybrid Cars: Today's Best Choice

Ferdinand Porsche designed a hybrid automobile in which a gasoline engine powered two electric motors that drove the front wheels of his "Lohner-Porsche" — a century ago, in 1899.

Hybrid automobiles have both a gasoline engine and an electric motor. Some hybrid prototypes are basically electric cars, where the gasoline engine comes on automatically when needed and operates at a constant, efficient, low-polluting speed purely to recharge the batteries. Other hybrids combine a small gasoline and a small electric motor to power the drive train together. Two hybrid vehicles, the Toyota Prius and the Honda Insight, are slated to hit U.S. showrooms in 2000.

Toyota Prius

The Prius' gasoline engine provides 58 horsepower, and its electric motor produces another 40 horsepower. An onboard computer decides automatically when each power source should be drawn upon. Working together, they make a very nimble little car.

The Prius already has been selling briskly for two years in Japan. The Japanese model gets 66 miles per gallon. *It can travel from New York City to Chicago—more than 800 miles— on a single tank of gasoline.* The American model has slightly more power, but still gets excellent mileage. It meets California's Super Ultra Low Emission Vehicle (SULEV) pollution standard—which will not even go into effect until 2004.

In the summer of 1999, Toyota loaned me the first U.S. prototype Prius to test-drive for two weeks. I put it through the wringer, from the Pacific slope rainforest to the high desert. I used it in daily commuting over the steep hills near my home in Seattle. I took it on a 500-mile swing over the Cascade Mountains and through the Yakima Reservation to the Columbia River Gorge without refueling.

The Prius accelerates swiftly, brakes crisply, lopes up mountains like a goat, and maintains good contact with the road through curves. Very comfortable for two passengers, it is as comfortable as any compact car for four. Its regenerative brakes recharge the batteries whenever you apply them. It has electric doors and windows, air conditioning, and a good sound system. Toyota's tentative marketing slogan for the Prius is "Energy Efficiency with No Compromises."

In the intelligently designed cockpit, everything is where it should be. You can turn on a nifty little liquid crystal screen that shows you graphically whether the gasoline engine and/or the electric motor are running, whether the batteries are being recharged or drawn down, what the external temperature is, and assorted other information.

Unlike the EV1 and the Honda Insight, the Prius does not

have a radically different shape. Its nose looks a lot like a new VW Beetle, and the rest of it looks like a Toyota Corolla. During two weeks of driving, only a handful of people noticed that I was driving anything special. (Those that did all asked me where I plugged it in. The answer is that I didn't. The nickel-metal hydride batteries are recharged solely by the engine and the brakes.)

American manufacturers all have prototype hybrids. But over late-night beers their executives confide that fuel cells are the real wave of the future. They would much prefer to plod along with yesterday's cars until fuel cells are ready to capture the market.

But as the computer industry shows us every month, there can be a huge market eager to buy products that are better than yesterday's machine, if tomorrow's perfect machine is not yet available. Ford, GM, and Chrysler all hope to sell fuel-cell vehicles by 2004 or 2005. If they miss that target by a few years, they may be handing Toyota an enormous number of car buyers for the interim, and a public relations coup to boot.

The Prius is the current state-of-the-art in production vehicle technology. Starting in mid-2000, it will be rolled out in city after city, as quickly as Toyota technicians are taught how to service the hybrid devices under the hood.

Honda Insight

Honda currently plans to introduce its own two-seat hybrid, the Insight, in the United States in early 2000. Powered by a 1.0-liter, three-cylinder engine with variable valve timing and an electric motor that runs off a nickel-metal hydride battery, the Insight expects to get *70 miles per gallon* on the highway in its U.S. configuration. Initial plans call for just 5,000 sales in the United States in 2000, but that could expand dramatically if demand is high. I was unable to preview an Insight at the time of this writing, but published reports are mostly favorable. (Its shape is reminiscent of the early Honda Civic CSX.) It accommodates just one passenger besides the driver, but that is sufficient for virtually all second cars in today's smaller families.

GM

General Motors is rumored to be planning to convert its two-passenger electric vehicle into a hybrid. No date has yet been announced for the introduction of a reconfigured hybrid EV1. However, GM and Toyota have agreed to pool all green-vehicle technologies, so Toyota's hybrid experience should allow GM to fast-track such a vehicle if it chooses to do so.

Race for the Market

Less than a year ago, I asked a very high-level U.S. auto executive when I could expect to buy a safe, comfortable, high-quality 55 mile-per-gallon vehicle. He joked, "You have at least a decade to save up for a down payment."

After a moment, he continued more seriously. "Unless there is a coup in Saudi Arabia. If that happens, I'll change my estimate to about one year."

If the Prius succeeds, I suspect the Big Three will suddenly find ways to produce a similar car by 2002, and that it will be affordable. Otherwise, they will watch Toyota and Honda eat their lunch. 🌎

Down the Road: Tomorrow's Cars

If your car gets 20 miles per gallon, you produce one pound of carbon dioxide for every mile that you drive.

A physicist's idealized concept of an efficient vehicle is one that operates without friction. At a steady speed on a level road, it would consume no energy. Energy used for acceleration would be recovered from braking. Energy used climbing hills would be recovered while descending. In the real world, of course, there is friction. Engine parts rub against one another, tires grip the road, and the chassis pushes through resisting air (as you learn when you stick your hand out the window when going 60 miles per hour). But automobile manufacturers could approach the physicist's ideal much more closely.

Environmentalists have long argued for lighter-weight, more aerodynamic cars, with innovative propulsion systems, more efficient tires, and better transmissions. Toward that end, the environmental movement has (unsuccessfully) sought for the last 15 years to ratchet the federal Corporate Average Fuel Efficiency (CAFE) standards upward.

The PNGV

Back in 1993, President Clinton rolled out his Partnership for a New Generation of Vehicles (PNGV). The PNGV combines tax-payers' money (through seven federal agencies), the scientific excellence of our national laboratories, and the engineering know-how of Detroit in an effort to design a car that uses less fuel. By 2004, the PNGV hopes U.S. manufacturers will be able to produce a U.S. vehicle that has roughly the same characteristics as the already-on-the-market Toyota Prius.

Actually, the most likely 2004 PNGV vehicles will be *inferior* to the Prius in one important regard: they will probably use diesel instead of gasoline engines. While diesel engines get somewhat better mileage, they produce more nitrogen oxides (which cause smog) and soot particulates (which harm human health) than gasoline engines do. Sadly and ironically, the cars produced by the decade-long, multiple-billion-dollar PNGV effort may be banned from California—the nation's largest automobile market—because they cause too much pollution.

Even worse, the auto companies have no obligation to actually *produce* any of these PNGV cars. Cynics think that the PNGV was simply a politically astute ten-year reprieve for the domestic auto industry from threats of higher Corporate Average Fuel Efficiency standards.

Fuel Cells

The standard internal combustion engines in today's cars burn gasoline to produce energy in the form of heat, which is used mechanically to turn a drive shaft. Fuel cells, which are discussed on page 41, combine hydrogen fuel with oxygen to produce energy in the form of electricity, which in turn runs an electric motor. Fuel-cell cars have all the advantages of electric cars, but, like hybrids, they make their electricity onboard instead of sucking it from a wall socket. Fuel cells and their attendant electric motors are extremely efficient. And when fuel cells are able to use pure hydrogen as a fuel, the electricity they generate will be pollution-free.

Fuel cells have been used for a long time in our space program, but were thought too large and costly for autos. Now they have been made lighter and cheaper. DaimlerChrysler and Ford Motor Company are collaborating with once-tiny Ballard Power Systems of Vancouver, British Columbia, to develop fuel-cell cars. Each plans to have five working prototypes on California roads in the near future, and for sale by 2004. DaimlerChrysler has a goal of selling at least 100,000 fuel cell vehicles by 2005. GM and Toyota also have teamed up to work on fuel cells, and Toyota announced that it plans to market a fuel cell vehicle by 2002.

Major hurdles remain. In late 1999, automotive fuel cells still cost ten times too much and were about 600 pounds too heavy for mass production for automobiles. However, fuel cells have few moving parts and are simpler than gasoline engines, so they should be affordable and reliable once they are mass-produced. The most important obstacle facing them may be the creation of neighborhood filling stations with hydrogen pumps. It is a chicken-and-egg problem.

As discussed earlier, hydrogen is typically obtained today by breaking down a fossil fuel to extract its hydrogen. During this process, the carbon in the gasoline, methanol, naphtha, or other fuel is emitted as a greenhouse gas. In other words, although the fuel cell *itself* runs clean—producing only water as a by-product—the hydrogen production process emits greenhouse gas. It is not yet clear whether this hydrogen extraction will be done aboard each vehicle (essentially requiring each car to carry a mini-refinery), or at service stations, where the car's tanks would be filled with hydrogen.

The ultimate goal is to produce hydrogen by using solar

energy to split water into hydrogen and oxygen. Then, when the fuel cell recombines hydrogen and oxygen to produce electricity, its only by-product will be clean water—completing the cycle. In essence, the energy in sunlight will be trapped to power automobiles.

For obvious reasons of self-interest, the oil industry vigorously opposes new super-efficient automobile motors that eliminate the need for gasoline. Oil-industry opposition will make the challenge of building a nationwide system of hydrogen filling stations for fuel cells much more difficult. Once one gasoline company breaks ranks, it will force its competitors to follow suit or risk losing share in an important new market. The CEOs of both BP and Shell have spoken of this as a likely development (albeit in the distant future). This, in turn, could trigger a chain of events that would lead to major investments in solar energy as a source of clean hydrogen.

Hypercars

Energy efficiency guru Amory Lovins has proposed a radical redesign of the automobile that he calls the hypercar. Lovins advocates advanced aerodynamic design for all surfaces, including the bottom; light, strong, composite materials; hybrid engines; regenerative braking; much-improved tires; and a host of other features. He argues that these off-the-shelf technologies can produce huge synergies when used in combination. (A lighter, more aerodynamic body needs a less-powerful engine and a lighter transmission. These, in turn, *also* weigh less, so an even smaller engine is adequate, etc.) Lovins makes a compelling case that we can mass-produce a safe, affordable, high-performance vehicle that gets over 180 miles per gallon.

The hypercar is a brilliant exposition of what technology, intelligently applied, could do to enhance the performance of the automobile. It is the car one might expect if Apple designers, Intel technologists, and Microsoft marketers collaborated to produce an automobile. In a sensible world, it is what we all would be driving. Check it out at *www.hypercarcenter.org.* 🌍

Beyond the Lonesome Driver: Transportation Alternatives

In many European cities, bicycles account for 20 to 30 percent of all trips.

Every time I see a lone driver in an eight-passenger SUV, I wonder whether the person at least considered all the other ways he could be getting to wherever he's going—without using such a large share of the Earth's resources.

Carpooling

Consider the simple mathematics of carpooling. If you drive alone in a car that gets 25 mpg, you are getting 25 passenger-miles per gallon. But if you carpool with a friend in the same vehicle, you will each average 50 passenger-miles per gallon. And if you add two more friends to your pool, you will each get 100 passenger-miles per gallon—better than any commercial car currently available.

Unfortunately, cars in the U.S. have been carrying fewer passengers every year. Although automobile fuel efficiency increased dramatically between 1972 and 1992 (before the SUV inundation), half those savings were negated by decreases in vehicle occupancy. Dagwood's carpool has become an anomaly. If current trends continue, by 2020 one out of every three cars won't have a driver.

Corporate Commuters

Employers often provide free parking as a tax-free benefit, thus subsidizing workers who drive cars. If that same amount of subsidy were available for other forms of commuting, it would help restore the balance. Employers know that congestion leads to a discontented work force. Preferential parking, subsidies, prizes, access to diamond lanes, sophisticated computer matching services, and other incentives are being used to increase the number of passengers in the typical commuter's car.

Buses and Subways

Public transit has the potential for greater benefits than most other transportation alternatives. But if we build it and they don't come, it can also be a great net loser. If the city operates a bus, or trolley, or subway, and few use it, we are consuming fuel for no purpose. A full bus expands the mathematics of carpooling many fold—full buses can get 1,000 passenger-miles per gallon. But an empty or nearly empty bus is a net loss; we would all have been better off if the two or three riders had driven cars or taken taxis.

Public transit offers huge potential benefits to our cities. Where subways have been in existence for decades, neighborhoods near subway stops have grown more dense. People travel efficiently between these stops and can easily walk to their destinations when they leave the system. It takes a while for such land-use patterns to evolve, but cities ultimately are reshaped into an efficient, attractive collection of urban villages.

The public transportation field is one where highly individualistic Americans may have much to learn from other cultures. Curitiba in Brazil has an extraordinarily successful bus system with a clever passenger-loading setup that is more efficient, comfortable, and popular than any bus system in the United States. Japan's subway systems are so regular that people set their watches by the cars' arrivals. Europe's coordination of the arrivals and departures of trains, ferries, trolleys, and buses has much to teach us about transportation planning, if we are not too proud to learn.

The Bicycle

The bicycle is the most elegant transportation device of all. It is even more energy efficient than walking! A bicyclist traveling at ten miles per hour uses about 100 BTUs per passenger-mile—as opposed to a pedestrian who uses about 500 BTUs per mile at 2.5

miles per hour. The bicyclist obtains the energy equivalent of a 1,000-mpg car—and consumes food, not oil. When I bike to work, I tell myself that I'm really saving time, because I don't have to take an hour off from work to go to the gym.

Forty percent of automobile trips in the United States are bike-sized (two miles or shorter). More than a quarter of all trips are less than a mile. The first mile is not only the easiest to bicycle, but also the most valuable to eliminate from the car, because 90 percent of the emissions from a seven-mile automobile trip are generated in the first mile, before the engine warms up.

The United States has 750 registered motor vehicles per every 1,000 persons (including toddlers and the aged). Europe averages 270. India has seven. China has eight. The global fleet of vehicles is increasing at the rate of about 16 million vehicles per year. Even so, three times as many bicycles are made each year as cars.

We should do a lot more to make life better for bicyclists: more bicycle trails and special lanes, more bike racks and lockers, and more bike carriers on buses. If you're interested in trying the bicycle route, start at your local bicycle repair shop. These shops often serve as clearinghouses for bicycle advocacy groups, which you can join. 🌍

ENERGY EXTRAS

Copenhagen provides 2,300 bicycles for public use. Businesses pay for the bikes in return for advertising space on them. Riders pay a refundable $3 deposit when they pick them up from bike stands. A Danish newspaper tracked one of these city bikes for 12 hours and found it spent only eight minutes at bike stands waiting for new users. Japan has 2,500,000 bicycle storage spaces in train depots.

If America were to make a serious commitment to bicycling, and this were reflected in the movies, television shows, and magazines that promote our lifestyles abroad, it could have an enormous effect on global automobile culture.

Move Closer to Work

According to a recent national study by the Texas Transportation Institute, traffic congestion cost travelers in urban areas 4.3 billion hours of delay, 6.6 billion gallons of wasted fuel consumed and $72 billion of time and fuel cost in 1997.

Americans drove 60 percent more total miles in their automobiles last year than the Germans, French, British, Japanese, Canadians, Mexicans, and Swedes combined. Per capita (including toddlers), the average American drives nearly 9,000 miles a year—more than three times the distance from New York to Los Angeles. If everyone in the world drove as many miles as we do, the total would be more than 51 trillion miles per year. Since the mean distance from the Earth to the moon is 240,000 miles, that is equal to 213 million trips to the moon every year.

When that volume of American driving is multiplied times the inefficiency of the cars in America's new vehicle fleet—a fleet that is more than half sport utility vehicles, vans, and light trucks this year—the gulf is even greater. That is why Ameri-

---

ENERGY EXTRAS

Marking the outer limits in terms of land use, Atlanta, Georgia is the fastest-growing city in history. In 1990, it stretched 65 miles from north to south. By 1999, it stretched 110 miles, and it is still growing. The resulting congestion has led Atlanta to be downgraded by many lists that used to rank it as one of the most desirable business locations in America.

---

cans, with 4.5 percent of the world's population, consume 43 percent of the world's gasoline.

How sustainable a ratio do you think that is?

How many wars will we fight to protect it?

Suburban commuters have little choice but to drive a lot, because suburbs are designed for driving. In the 1970s and 1980s, a 1-percent increase in population led to a 6- to 12-percent increase in land consumption. Between 1970 and 1990, the population of Greater Chicago grew just 4 percent, but the physical metropolitan area grew 50 percent. In the 1990s, we often saw a 10- to 20-percent increase in land use for each percent of population growth.

According to High Mileage Moms, a study by the Surface Transportation Policy Project, suburban mothers make an average of five car trips a day getting to work, running errands, hauling children and elderly parents, etc. Single suburban mothers spend 75 minutes a day behind the wheel, while married mothers drive an average of 66 minutes. In the last decade, the number of children who walk to school has declined by 23 percent. People who moved to the suburbs so that their children would have lawns to play on are finding themselves with little time to play with their kids. Commuting accidents are epidemic; those who moved to the suburbs to avoid crime are more likely to be injured in a commuting accident than a city resident is of being mugged.

Commuter congestion has emerged as a vibrant political

issue in much of the United States. In the November 1998 elections, voters approved 72 percent of the 240 state and locaballot measures to limit growth or preserve open space.

It seems almost too obvious to say that living close to where you work reduces your contribution to global warming. If you live within walking or bicycling distance of work, you can reduce the global warming impact of your commute to zero. 🌍

Work Where You Live

In 1990, there were 3.6 million telecommuters in the U.S.

Why wait for casual Fridays? If you work at home, you can work in your running shorts every day. When you telecommute, it makes little difference how far away your home is from your company. It is as easy to send an electron halfway around the world as it is to send it next door.

Telecommuting is a solution to air quality and traffic congestion; it improves public health, stimulates productivity, fuels economic growth, and enhances your quality of life. Imagine how much more energy-efficient it is to send your work over a wire than it is to put your body into two tons of metal and haul it into the office. Teleconferencing, video conferencing, and Web commerce all save time and energy.

> **T O O L B O X**
>
> *For more information on telecommuting, visit the Web site of the International Telework Association & Council,* www.telecommute.org; *or visit consultant Gil Gordon's site at* www.gilgordon.com.

The number of American employees who telecommute from their homes jumped to 10 percent of U.S. adults in 1999, when more than 19.6 million people reported telecommuting at least occasionally. Major employers such as AT&T, Hewlett-Packard, and Merrill Lynch allow some employees to commute via computer rather than car. Other employers are becoming more flexible about what defines a workday. By working four ten-hour days, or by working from home one day a week, you become part of the pollution solution. And over a year, that's 50 fewer days that you waste sitting in traffic.

Some jobs lend themselves to telecommuting and others don't. Lathe operators and short-order cooks will never be able to telecommute. But an ever-increasing fraction of Americans are "knowledge workers" rather than manual laborers, so opportunities for telecommuting should continue to grow.

Your employer can also save money and energy by using less office space. Office buildings are the fastest-growing use of electricity in the U.S. Each employee who teleworks can save his or her employer $10,006 in reduced absenteeism and job retention costs, according to research by the International Telework Association & Council. That means a company with a 100 employees, 20 of whom telework, could potentially realize a savings of $200,000 annually. I typically work at home one or two days a week, when I have a project that requires concentration and I need to avoid constant interruptions. Telecommuting works best if you're a self-starter with motivation and discipline. If this sounds like you, propose a plan to your company, spelling out costs and benefits discussed above. 🌍

Trains and Airplanes

Commercial aviation uses 13 percent of total transportation energy in the United States, while trains account for just 2 percent.

Aircraft emissions have been doubling every decade, despite enormous increases in efficiency. The amount of fuel consumed per passenger mile has dropped more than 50 percent since 1970, partly because planes are mechanically more efficient, and partly because each plane carries more passengers. However, far more people are flying far more miles, so total emissions are soaring.

As much as 10 percent of America's total greenhouse gas emission from all sources may come from airplanes—half as much as comes from road transport. That is because airplanes produce not just CO_2, but also nitrogen oxides that are converted to ozone—a greenhouse gas—in the troposphere. (Ozone is desirable in the stratosphere, the layer in the atmosphere above the troposphere, because it blocks harmful UV radiation. However, it is harmful in the troposphere, the lowest layer of the atmosphere, where it contributes to global warming.) In a greenhouse-constrained world, airplane manufacturers will have to find a way to solve the tropospheric ozone problem.

The most energy-efficient way to move both people and goods between cities is by rail. In fact, that is how it is done throughout much of the world. In Japan, Germany, France, and Italy, high-speed electric trains running on welded rails link hundreds of cities. By picking up and delivering passengers right from the heart of a city's business district, trains eliminate the time, expense, and CO_2 associated with taxi rides from ever-more-distant airports. Destination-to-destination, high-speed trains are faster than planes for most trips of 400 miles or less.

Trains are a natural market for fuel cells. Today's diesel-electric locomotives are essentially small electric power plants that drive 4,000 horsepower electric motors. Trains have more space to store hydrogen or methanol, and they would make an ideal vehicle to test medium-sized fuel cells.

I confess that I have a small child's love of trains. I have taken the Trans-Siberian Express across the full breadth of Asia; the Orient Express across Europe; the Indian-Pacific across Australia. But for decades, I avoided American trains whenever possible, as I had uniformly bad experiences whenever I relented. Our railroads tried hard to divorce themselves from the passenger business to just the hauling of freight.

After Amtrak—a public corporation with public subsidies and

<table>
<tr><td>

ENERGY EXTRAS

Amtrak has begun introducing 20 high-speed "Acela Express" trains on the East Coast. Traveling at speeds up to 150 miles per hour, they shave about two hours from the Boston to New York run. The Acela has a surge-protected power outlet for laptops by each row of seats, two channels of audio, luggage bins twice the size of airliner bins, and beer on tap in the lounge car.

</td></tr>
</table>

oversight, and all the surly, bureaucratic incompetence that that can imply—took over the intercity passenger business, things initially got even worse. As a result, today's trains haul an awesome 1.5 billion tons of freight each year, but they account for scarcely 1 percent of the passenger miles logged by airlines.

In the last few years, however, an extraordinary new spirit of customer service has emerged at Amtrak. It is one of the great, unheralded turn-around stories in American industry. Many train aficionados are daring to hope that passenger trains may again become competitive.

My wife and I recently took Amtrak's Coast Starlight from Seattle to Los Angeles. It's a 35-hour trip, so we paid for a bed. Here's what we learned:

• The journey by railroad cost twice as much as the cheapest, 14-day-in-advance airfare, but was comparable to a standard, booked-months-in-advance airfare.

• The trip took ten times as long by train as by plane.

• On the other hand, airlines offer a tasteless box lunch whereas Amtrak gave us six superb meals, which were included in the price of our ticket. They even gave us free souvenir mugs.

• The train was spacious and relaxing, and the theater car showed first-run movies without cheap earphones.

• My Amtrak compartment had a comfortable desk with an electrical outlet for my laptop.

• Airplanes are jammed with tense, cranky people; every train traveler I encountered was relaxed and polite.

• Airline personnel are often pressured and disagreeable. Amtrak employees were uniformly pleasant and eager-to-serve. (When I commented on this, one employee said, "We aim to be the Nordstrom of transportation.")

• Airports are hard to get to and are filled with long lines. Train depots are right downtown, and they are increasingly recapturing their earlier architectural grandeur.

If time is an issue, America's slow trains will not work for you on trips of more than a couple hundred miles. But if you want to enjoy the journey, and cherish the opportunity for relaxed hours of reading or of uninterrupted work (I logged perhaps 20 hours of work on this book en route to Los Angeles), the train is a hands-down winner. 🌎

Clean Homes
Shelter from the Storm: Efficient Buildings

The average home contributes twice as much greenhouse gas pollution as the average car.

Temperatures on the surface of the Earth vary about 270°F, but our bodies cannot endure more than a 10° F fluctuation in blood temperature for long. So while birds migrate and bears hibernate, humans build shelters against the heat and the cold. Until the post-World War II boom, our homes were generally oriented with the sun, prevailing winds, and local terrain in mind. They employed appropriate building materials that either stored heat or blocked it. The ancient Greeks, Babylonians, and Anasazi also did this, not because of building laws or zoning codes, but because it was just common sense.

As we enter the new millennium, most buildings have little or no relationship to nature. Whether in Texas or Maine, most new buildings look very much like one another. Moreover, most buildings are much the same on all four sides, ignoring the potential power of the southern sun.

The only consideration architects of many modern buildings give to the environment (other than by meeting the minimal requirements of building codes) is in the size of the huge furnaces and air conditioners they specify. Individual dwellings are even less energy efficient than most large structures. Some of the emissions go up the chimney from the house's furnace. Others go up the smokestack of the power plant that provides its electricity, an indirect but nevertheless harmful result of our inefficient homes. Everything considered, the average home produces as much carbon dioxide as two cars.

The last several years have seen dramatic improvements in the efficiency of many major appliances—improvements forced by federal efficiency standards that are the fruit of a strong environmental movement working collaboratively with a handful of enlightened appliance manufacturers. (Unlike automobile manufacturers, who all uniformly oppose fuel-efficiency standards, some high-quality appliance and lighting manufacturers have welcomed them.) Nevertheless, Americans are still using more energy in their homes every year. How can this be?

Many new furnaces are 95 percent efficient, whereas most old furnaces are less than 50 percent efficient. But people tend to replace things only when they break. Furnaces and appliances often last from ten to 30 years, so it takes a long time for efficient new models to penetrate the whole market.

Another part of the explanation is that the average new American home has been getting bigger and bigger (2,100 square feet today), even though family size has been shrinking (from 3.6 to three people).

Improving the Efficiency of Your Home

So much for the statistics. You don't live in a statistic. You live in your home. While it is cheaper and easier to switch to a low-emission car than to a low-emission house, there are cost-effective ways to dramatically reduce the emissions attributable to your current household. Some of these reductions will come from more efficient furnaces, more efficient appliances, more efficient lighting, etc., which I'll discuss in subsequent chapters. Let's focus first on the flows of energy through the structure itself.

The goal of an energy-efficient building is to minimize the transfer of heat between the inside and the outside. Heat passes in and out of buildings in three ways: conduction, convection, and radiation. Each one must be managed differently.

Conduction refers to heat transfer in solids. If you stick the business end of a silver spoon into a hot cup of coffee, the handle soon gets hot. It was warmed through conduction. The same is true for the walls of your home. The principal way to combat conductive losses is insulation. Insulation is material that conducts heat very poorly. Insulating your home is similar to using a thermos instead of a glass jar—insulation keeps the stuff inside warm while it keeps the outside air cool. Heat loss by conduction is inversely proportional to the thickness of the conducting material, so a good rule of thumb is that doubling the

> **ENERGY EXTRAS**
>
> *Many Republicans oppose all efficiency standards. They try to gut every law that authorizes the government to set a standard; if they fail, they then try to kill any bill that appropriates money for standard-setting. Sometimes they win; sometimes they lose. As a result, there is a crazy-quilt quality to appliance efficiency, with regulated appliances being far more efficient than unregulated ones.*

thickness of the insulation cuts the potential heat loss by half.

The most important part of the house to insulate is the attic, but every surface can lose heat through conduction. If your basement is unheated, or if you have a crawl space, insulate the underside of your floor.

There are many types of insulation. The federal Energy Star program has a good Web site to help you pick the right material and decide how much to install for your climate at *www.energystar.gov.* (Most insulation retailers will give you an Energy Star brochure containing this same information, though you may have to ask for it.) You may want to install more than the federal recommendation if you are more interested in saving

---

T O O L B O X

Energy Star is a U.S. program run jointly by the EPA and the U.S. Department of Energy. Unlike appliance efficiency standards, Energy Star is purely voluntary. Companies that choose to participate obtain the right to use the Energy Star logo only if their product is significantly more efficient than required by government standards.

Look for the star on home appliances, home electronics, office equipment, heating and cooling equipment, lamps, and other major energy-using devices. Some local utilities offer rebates for the purchase of Energy Star-rated appliances.

You can visit the Energy Star program's frequently updated Web site (www.energystar.gov) to get information on specific products, manufacturers, and stores that carry the products. Or call the toll-free hotline at 1-888-STAR YES.

---

energy than in saving money. Everyone should install at least as much as Energy Star recommends.

Convection occurs when air circulates between the inside of your house and the great outdoors. Pressure differences, caused by wind or by your heating system, force heated air out of the house and replace it with outside air. In most older houses, half of all inside air goes outside every hour, mostly through very small openings. The amount of heat lost through a quarter-inch crack along a three-foot attic door can cost you more than 20 gallons of fuel oil in a moderate winter. Weather stripping, caulking, and storm doors and windows can help reduce unwanted heat loss through convection.

Radiation is the type of heat you feel when you are standing near a hot wood stove. "Radiant barriers" is a fancy name for thin plastic sheets coated with a shiny surface (such as an extremely thin layer of aluminum) to reflect back radiant energy. They are typically stapled to attic rafters or sloped trusses inside a house, shiny side down, to reflect radiant heat. Some rolls of insulation have the radiant barrier built right in.

But to reduce radiant heat loss, the best place to look in most houses are the windows.

Look to Your Windows

People and objects at room temperature radiate energy directly through windows to a colder environment outside. The cold on the other side of the window actually sucks out your body heat. That is why you feel chilly standing near a single-pane window in cold weather, even if the air around you is toasty. Radiant sunlight also has its greatest impact on windows. Sunshine is the boon of solar-designed houses and the bane of many others.

When your house is too hot inside, unwanted solar radiation can be screened by awnings, shutters, or insulating shades; by reflective films; and by external shade trees. Even white curtains will go a long way in reflecting back unwanted sunlight. A *low-e* coating is like a radiant barrier for windows, reducing heat loss in winter and heat gain in summer. Windows also transfer heat through conduction and convection. Confusingly, windows are often rated by either R-value (resistance to the

TOOLBOX

Hurd, Pella, and Visionwall all make superwindows using Heat Mirror™. Viking makes a less expensive window with comparable thermal characteristics using a triple-pane window with low-e glass. Superwindows make the most sense in very cold regions. (In hot climates you are better off shading the window to stop radiant gains.) In cold, sunny Aspen, Colorado, a superwindow can provide a net heat gain even on the north side of the house. More solar energy reflects off the snow through the superwindow than heat energy escapes back out!

flow of heat) or U-value (ability to transfer heat). You want *high* R-values or low U-values. R-2 is the same as U-0.5, R-4 is the same as U-0.25, and so on.

Until the 1990s, most windows were rated on their R- or U-value at the center-of-glass. But there is often significant heat loss through the edge of the glass and the frame, so a more useful measure is the whole-unit value. A typical, well-sealed, double-pane window has a whole-unit R-value of 2.0. During a mild winter, an average single-pane window loses as much energy by conduction as is produced by a 75-watt light bulb burning continuously. Minimize such losses by insulating your windows. Use double or triple glazing and tight, energy-conserving frames. Add storm windows where necessary. Movable insulating products, like window quilts, can also help.

Superwindows combine multiple layers of low-emissivity glass, excellent edge seals, insulated frames, and an inert, harmless argon or krypton gas between the panes (krypton is better). Heat Mirror™ can have a whole-unit rating up to R-6. Such a window allows only half as much heat transfer as the same thickness of fiberglass insulation!

In recent decades, we have developed techniques to change

the reflectivity of windows (much as some sunglasses adjust to the amount of sunlight). Windows can be designed to respond to the intensity of sunlight, to the temperature, or to human control (e.g., using a trivial electric current to turn a window into a one-way mirror). Although all these techniques have been technically demonstrated and have generated a lot of gee-whiz press, they remain too expensive to see general application by the consumer market. However, if oil prices double, such windows could become commonplace. 🐾

Heating and Cooling

Your home spends more of its energy allowance on heating and cooling than anything else.

Furnaces

Furnaces have become very fancy in recent years, with features like variable-speed blowers that can contribute to your comfort. In terms of fuel use (and hence emissions of carbon dioxide), four considerations dwarf all others: efficiency, correct sizing, a competent contractor who will cooperate with your desire to save energy, and a clock thermostat.

Efficiency. The standard measure of furnace efficiency is the annual-fuel-utilization-efficiency (AFUE) rating. The AFUE is expressed as a percentage of the energy in the fuel that is delivered as heat by the furnace. The higher the AFUE, the less fuel you will need to heat your home, and the softer your impact will be on the environment. Twenty years ago, the average furnace was about 65 percent efficient. Today, all new furnaces are required by law to be at least 78 percent efficient. The most efficient models have an AFUE of 97 percent—which is about as efficient as a furnace can hope to get. To retain your gas furnace's efficiency, get it tuned up every three years; an oil furnace should be maintained every year.

Size. A furnace that is too small will not adequately heat your house in the coldest months, no matter how hard it tries. To avoid this possibility, many contractors sell furnaces that are larger than necessary. But a furnace that is too powerful will cycle on and off more often than a correctly sized furnace, wasting energy, wearing out components more rapidly, and causing the temperature to fluctuate. Large furnaces also commonly require larger, more expensive ducts. Too large a furnace will cancel out the benefits you can obtain through ample insulation, efficient windows, and improvements to your house's shell. Standard calculations involving the house's size, design, insulation, etc.; the AFUE of the furnace; and the region's climate will allow your contractor to size your furnace properly. Make it clear that you do not want it oversized.

Gas boilers—both steam and hot water—vary more in efficiency, from 95 percent for the most efficient hot water systems (made by Dunkirk) to 81 percent for the least-efficient steam units (made by lots of manufacturers).

Oil furnaces are typically about 10 percent less efficient than gas furnaces (about 85 percent efficient instead of 95 percent efficient), but oil boilers and gas boilers are comparable.

For information about scores of different furnace models, see the ACEEE Web site (*http://aceee.org/consumer-guide*) or the *Consumer Guide to Home Energy Savings, Seventh Edition* (Washington, D.C.: American Council for an Energy-Efficient Economy, 1999).

T O O L B O X

If you want a complete energy makeover, check out this Web site: HomeEnergySaver.lbl.gov. By plugging in your zip code and some other pertinent information about your home and its heating and cooling systems, appliances, age, number of windows, etc., the site can calculate how much you're spending currently on energy. It then suggests changes you can make and shows you how much energy and money you'd save.

Natural gas furnaces of all sizes are now extremely efficient. Their annual-fuel-utilization-efficiency (AFUE) ratings are all between 94.0 and 96.9. You can't go wrong. All other things being equal between two models, choose the one that uses less electricity for its fan. The differences can be far larger than you would imagine. For example, for large furnaces, a Sears model and a York model have the same 94.0 AFUE, but the Sears model uses 242 kWh per year, while the York model uses 1,493 kWh per year. The York model could cost you an additional $125 annually to operate. Moreover, if your electricity is produced from coal, it would cause the emission of an extra one and a half tons of carbon dioxide every year.

A competent contractor. Selecting, installing, and maintaining a furnace are beyond the skill levels of most people who don't do this for a living. This is especially true of today's sophisticated, super-efficient furnaces. *Your most important furnace decision is in selecting the right contractor.* A 1999 study by the American Council for an Energy-Efficient Economy demonstrated that improved residential installation practices for heating and cooling devices could save an average of 24 percent of energy use in existing homes and 35 percent in new construction.

So look for someone who understands that saving as much energy as possible is of genuine importance to you, and who takes your concerns seriously. Make sure that the technicians who will actually do the work at your home are all trained and certified. Improper installation, incorrect venting, or sloppy duct work can cause you years of cost and aggravation. (The U.S. Department of Energy estimates that 20 to 40 percent of the heat produced in most furnaces is lost in duct systems.)

Recommendations from satisfied friends, family, and

co-workers are your best source. The local utility may also have a pre-screened list. If you choose a contractor from the yellow pages, check references carefully, and look for complaints with the Better Business Bureau. Insist on a contractor who takes pride in his or her craftsmanship.

Clock thermostat. A set-back thermostat allows you to set a lower temperature while you sleep and during times when everyone is out of the house. You can do this manually, of course, but we tend to forget. Mechanical thermostats are cheap and easy to set, but limited in their functions. Electronic models are more complicated but more versatile, allowing you to program different cycles on different days of the week. For electromechanical thermostats, *Consumer Reports* recommends the White Rodgers 7901 and the Honeywell CT1501. For electronic models, the Honeywell MagicStat 33, the Honeywell 34, and the Sears Weekender 91112 are recommended.

Cooling

Energy-efficient landscaping can shade your house and channel breezes to where they are needed. A white roof can reflect up to 70 percent of the sunlight that strikes it, whereas black shingles reflect only 5 percent. Attic insulation, radiant barriers, and adequate attic ventilation will all stop unwanted heat from entering your house.

In some climates, these measures are sufficient to ensure that a house does not overheat. In Seattle, where I live, few houses have air conditioning. We can get by with fans, which use far less energy and hence produce less greenhouse gas. In much of the country, however, air conditioning has become commonplace, and in desert regions it often uses more of the household energy budget than heating.

The efficiency of central air conditioners is rated according to their seasonal energy efficiency ratio (SEER). The SEER is the seasonal cooling output (in BTUs), divided by the seasonal energy input (measured in watt-hours). Older units typically have a SEER rating of about 6. By 1988, the SEER average had risen to 9. Today, air conditioners are required to have a minimum SEER of 10, and an Energy Star label indicates a SEER of 12 or higher.

For room air conditioners, efficiency is measured as an energy efficiency rating (EER). The EER is the ratio of cooling output (in BTUs) divided by the power consumption (in watt-hours). New federal standards scheduled to go into effect in October 2000 will require a minimum EER of 10, an efficiency level already met by many models.

If shopping today for a central air conditioner, look for a SEER rating of 12 or more. Central air conditioners qualify for an Energy Star listing if they have a SEER of 12 or more. Dozens of models have SEERs of 15 to 18. The most efficient room air conditioners have EERs of from 10.0 to 11.7 (the higher the better). For a list of the most efficient models of all standard sizes, see the ACEEE Web site (*http://aceee.org/consumerguide*) or the *Consumer Guide to Home Energy Savings, Seventh Edition* (Washington, D.C.: American Council for an Energy–Efficient Economy, 1999).

Although larger units sometimes have higher efficiencies, don't buy an oversized unit just to get the higher efficiency. An oversized unit will cool the room quickly, but it will remove only part of the humidity, leaving a damp, clammy feeling in the air, because not enough air has been circulated through the machine for the water to have been squeezed out of it. A correctly-sized unit will provide a higher measure of comfort and ultimately save energy.

When buying a central air conditioner, you should insist on the same due diligence by your contractor with regard to all the factors to be considered as with a furnace. Do not consider using a contractor who merely measures the square footage of your house. A good

> T O O L B O X
>
> *For each degree Fahrenheit you raise your thermostat in summer, when you are air conditioning, you cut roughly 5 percent off your energy use. For each degree you lower your thermostat in the winter, you cut your energy use by about 3 percent.*

contractor will poke around measuring insulation levels, checking radiant barriers, and examining for an accurate and sophisticated sizing calculation.

Heat Pumps are Cool...or Hot

Think of a heat pump as an air conditioner that can run in reverse. Like refrigerators or air conditioners, heat pumps use electricity to move low-grade heat in a direction in which it would not ordinarily flow. In the summer, they transfer heat out of a house into a warmer external environment, and in the winter they transfer heat into the house from a cooler external environment. Even as a refrigerator removes heat from the cool air inside itself and transfers it to the warmer kitchen, a heat pump can remove heat from the cool outside air and dump it into the warmer house.

Heat pumps can be very efficient, but they run on electricity and much energy is lost in conventional power plants when electricity is produced. Also, heat pumps are much less efficient as heat sources when outside temperatures get very low. (Ground-source heat pumps are less adversely affected by plunging temperatures than air-to-air heat pumps, but are typically twice as expensive and require extensive excavation to lay underground pipes.)

When shopping for heat pumps, look for one that has a SEER of at least 14 when used in a cooling mode, and a coefficient of performance of at least 3 in the heating mode. To qualify for an Energy Star label, an air source heat pump must have a SEER of at least 12. A huge number of models qualify.

For information about dozens of air source and ground source heat pumps of all sizes with the very highest efficiencies, visit the ACEEE Web site (*http://aceee.org/consumerguide*) or the *Consumer Guide to Home Energy Savings, Seventh Edition* (Washington, D.C.: American Council for an Energy–Efficient Economy, 1999).

In terms of global warming benefits, heat pumps make the most sense in regions with mild winters and hot summers, and where consumers can choose to buy electricity made from green sources. 🌎

Clean Energy Checklist

If the average American were to equip his home with only products that have the Energy Star label, he'd cut his energy bills—and his greenhouse gas emissions—by about 30 percent.

The American Council for an Energy-Efficient Economy has prepared a checklist of things that homeowners can do to save energy. The following suggestions, based on that list, provide a roadmap for making dramatic, economically-sound reductions in the amount of carbon dioxide your home produces.

To Do Today

• Turn down the temperature of your water heater to the warm setting (120°F). You'll not only save energy, you'll avoid scalding your hands. (For more information, see the following section on Hot Water.)

• Check to see whether your water heater has an insulating blanket. An insulating blanket can pay for itself in one year or less! (For more information, see the following section on Hot Water.)

• Start using energy-saving settings on refrigerators, dishwashers, washing machines, and clothes dryers. Check the age and condition of your major appliances, especially the refrigerator. You may want to replace it with a more energy-efficient model before it dies. (For more information, see the sections dealing with various appliances on pages 100-108.)

• Survey your incandescent lights for opportunities to replace them with compact fluorescents. These new lamps can save three-quarters of the electricity used by incandescents. The best targets are 60- to 100-watt bulbs used several hours a day. (For more information, see the section on Smart Lights.)

• Clean or replace the filters in your furnace, air-conditioner, and heat-pump.

This Week

• Rope-caulk leaky windows.

• Visit the hardware store and buy a water-heater insulating blanket, low-flow showerheads, faucet aerators, and compact fluorescents, as needed.

• Assess your heating and cooling systems. Determine whether replacements are justified, or whether you can retrofit them to make them work more efficiently.

This Month

• Collect your utility bills. Separate electricity and fuel bills. Target the biggest bill for energy conservation remedies.

• Go into your attic or crawlspace and inspect for insulation. Is there at least as much as is recommended by Energy Star for your climate? (Check out *www.energy star.gov* or call 1-800-STAR-YES.)

• Insulate hot-water pipes and ducts wherever they run through unheated areas.

• Seal up the largest air leaks in your house, the ones that whistle on windy days or feel drafty. The worst culprits are usually not windows and doors, but utility cut-throughs for pipes, gaps around chimneys and recessed lights in insulated ceilings, and unfinished spaces behind cupboards and closets.

• For expert advice on your home as a whole, schedule an energy audit (ask your utility company or state energy office for recommendations). Try to find an auditor with a blower door, which creates a high-pressure area inside your house to help locate the worst air leaks. All the little, invisible cracks and holes may add up to as much as a wide-open window or door, without your ever knowing it.

• Install a clock thermostat to set your house's temperature back automatically at night and when you're not home.

This Year

• Insulate! If your walls aren't insulated, have an insulation contractor blow cellulose into them. Bring your attic insulation level up to adequate levels for your climate.

• Replace aging, inefficient appliances. Even if the appliance may have a few useful years left, replacing it with a top-efficiency model is generally a good investment for your personal pocketbook as well as the planet's health.

• Upgrade leaky windows with energy-efficient models or at least boost their efficiency with weather-stripping and storm windows.

• Reduce your air conditioning costs by planting shade trees and shrubs around your house, especially on the western and southern sides. 🌍

Hot Water

An energy-efficient showerhead can save an average of $0.27 each day on water bills and $0.51 on electricity.

Water heaters are the second largest energy users in the home, typically using about 20 percent of all household energy. Only space heating and cooling ranks higher.

What to Look For

Gas-fired water heaters are more efficient than electric heaters, because they deliver heat right where it is needed instead of losing most of it at a distant power plant. Electricity is a marvelous form of energy, capable of running computers and powering cellular phones, but it makes no sense to use electrical resistance heaters to warm up water.

If a gas connection is not available in your neighborhood, a heat pump is often a smart choice. Heat pumps reduce the electricity required to heat water by one-half to two-thirds—even more in very hot climates. In colder climates, ground-source heat pumps are more effective, but they require extensive excavation and make the most sense in new construction.

Solar water heaters produce no greenhouse gases at all. I've lived comfortably with stand alone solar water heaters in Hawaii and in the Sun Belt. Elsewhere they are usually backed up with a supplementary energy source.

Tankless ("on-demand") water heaters are especially good for vacation homes or when used in combination with solar water heating. They heat up water just as it is used,

> **T O O L B O X**
>
> *Excellent pamphlets available from the Florida Solar Energy Center, 300 State Road 401, Cape Canaveral, Florida 32920, 407-783-0300 can help you decide whether a solar water heater is right for you.*

so there is no loss of heat from a tank of hot water. Moreover, they are usually located right where the hot water is needed, so there are no pipe losses. They have a limited flow, however, and can generally be used only one faucet at a time. (Some tankless units also are so noisy you might as well be sharing your bathroom with a small jet plane. Listen before you buy.)

How to Use Them

Lower the temperature. The most important step you can take to save energy when heating water is to make sure you are not overheating it. For each 10° reduction you save roughly 13 percent on your energy bill for water heating. Most households are comfortable with water heated to 115-120° F, but most tanks go up to 140°. If you find you must dilute your hot water tap with cold water, the thermostat on your heater is probably set too high.

Insulate your heater. Insulating your water heater decreases the energy used to heat it up. Wrap your water heater in an insulation jacket, available at hardware stores and sometimes given away free with a new water heater, so that it is insulated to at least R-24 (only a few new ones come with enough insulation already installed). Tank blankets generally pay for themselves in less than a year. This is one of the more cost-effective things you can do in 15 minutes to save energy.

Insulate your hot water pipes. Use the split-foam-rubber pipe insulation available at any hardware store. Be sure to choose the right size for your pipes. Put it on with the crack facing down, and tape the seam with acrylic tape. Insulate hot water pipes wherever they are accessible, especially the first three feet from the water heater.

Install high-efficiency showerheads. High-efficiency showerheads can use as little as a sixth as much water as standard fixtures. They come in both aerated and non-aerated models. These devices should not be confused with cheap flow restrictors, which reduce the water flow but produce anemic sprays. High-efficiency showerheads are designed to produce excellent showers using less water.

Install aerators. Aerators can be as helpful on faucets as on showerheads. On a typical faucet, they can reduce the flow of water from 50 to 90 percent. Because aerated water thoroughly wets objects instead of bouncing off, these devices are effective even though they use far less water. Retrofitting one shower-head and two faucets at a total cost of $25 can—according to the Rocky Mountain Institute—reduce your annual electric bill by $86 (or your gas bill by $36, if you have a gas water heater). Depending on your energy source, you will reduce your annual carbon dioxide emissions by anywhere from 580 to 3,200 pounds. 🌏

Smart Lights

Roughly a quarter of the total electricity generated in the U.S.—$37 billion worth—is used for lighting. This is more electricity than the continents of South America and Africa, combined, use for all purposes.

Lighting is the electrical use most in need of improvement. Most lights are really heating devices that also provide some illumination. Even the most efficient fluorescent lamps convert only about 20 percent of the electricity they use into light; 80 percent is given off as heat. Incandescent lamps, the kind most people still use at home, are only one-third to one-fourth as efficient as fluorescents. Only 6 percent of the electricity used by incandescent lamps is turned into light; the rest is radiated as heat.

Many major office buildings produce so much heat from their lighting systems that they have no other heating system; some actually have to run their air-conditioners throughout the winter. In air-conditioned buildings, every two watts used for unnecessary lighting requires one additional watt for cooling.

The point is to use the most efficient type of lamp that is effective for a given job—in other words, to provide as much light as is required, and no more. Joseph Swidler, a former chair of the Public Service Commission of New York, once remarked that the hallway outside his office had "more than enough light for fine needlework, miniature painting, or engraving counterfeit money," though it was only used for walking from one office to another.

There are many types of lamps to choose from, including standard incandescent, compact fluorescent, tube fluorescent, tungsten halogen, metal halide, mercury vapor, high-pressure sodium, and low-pressure sodium. All of these, and many others, have particular applications where they make sense. The first four are of most relevance in the house.

Standard Incandescent Lamps

Thomas Edison's first successful lamp, in 1879, was an incandescent. Modern incandescent light is produced by passing an electric current through a tiny coil of wire made out of tungsten, a metal that glows when it is heated. The incandescent lamp is one of the great success stories in industry. Today, it dominates literally all household lighting applications. However, since it converts only about one-twentieth of the electricity it uses into light, it is also glaringly inefficient.

Higher-wattage incandescent bulbs are more efficient than low-wattage bulbs. A 100-watt bulb uses twice as much electricity as a 50-watt bulb, but provides more than twice as much light. (Of course, don't use a 100-watt bulb if a single 50-watt bulb provides all the light you need.)

Incandescent lamps have the shortest lives of any common bulb. Some manufacturer have begun marketing "long-life incandescent bulbs, which have thicker filaments that don't burn out as quickly. These are even less energy-efficient  than standard incandescents, so they don't help combat global warming.

Incandescent lamps make the most sense in places where they are seldom or briefly used. They may be the best choice for a guest room, for the furnace room, or for stairways when instant-on lighting is a safety requirement. For longer-burning lamps, you'll save money and avoid more CO_2 emissions by using a more efficient light bulb.

Compact Fluorescent Lamps

The next time you stay in a nice hotel, peek under the lampshade. It's likely you will find a compact fluorescent lamp (CFL). Hotels are keenly aware of the cost of lighting. At the same time, they don't want guests complaining about dim 40-watt bulbs they can't read by. CFLs are the answer: 25-watt lamps that burn as brightly as 100-watt incandescents. Marriott Corporation has saved millions of dollars by switching to such bulbs.

CFLs last ten times as long as standard incandescent bulbs and use only a quarter of the energy to produce the same amount of light. No longer heavy, clunky devices that don't fit your fixtures, modern CFLs with electronic ballasts come in many shapes—short tubes, hoops, loops, and corkscrews—and will fit most fixtures. Although CFLs cost substantially more than incandescent lamps at the cash register, incandescent

---

T O O L B O X

Green Lights is a voluntary program run by the U.S. Environmental Protection Agency. It encourages corporations to sign an agreement to adopt cost-effective lighting retrofits. Participants reduce their lighting bills dramatically while maintaining lighting quality. They often achieve annual rates of return on their investment of 30 percent or more.

If all nonresidential space in the U.S. met Green Lights' standards, we would reduce our total use of electricity by about 12 percent. For more information visit their Web site (www.epa.gov/GCDOAR/GreenLights.html) *or call* 202-775-6650.

---

bulbs will cost you twice as much as CFLs in the long run.

The math is simple. Suppose you pay $20, the full retail price for a top-of-the-line 25-watt CFL. (You can often get *very* substantial discounts—up to 50 percent—at warehouse stores.) The CFL bulb will last 10,000 hours. A standard 100-watt incandescent that produces the same amount of light will cost perhaps $0.50, and last up to 1,000 hours. You will spend $5 for the ten incandescent bulbs needed for 10,000 hours of light. I know that's still much cheaper than the CFLs, but you haven't added in the energy cost yet.

Using the national average electricity rate of 8.4 cents per kilowatt-hour, you will pay $84 to operate the 100-watt incandescents for 10,000 hours. Add the $5 for ten incandescent bulbs, and your total cost is $89. The 25-watt CFL will need $21 of electricity to shine for 10,000 hours; add the $20 purchase price, and the total is $41. The incandescents cost more than twice as much.

Nine companies market CFLs that qualify for the Energy Star label.

Alphabetically, they are:

- AmericanPower Products
- Feit Electric Co.
- GE Lighting
- Lights of America
- MaxLite-SK America
- Osram Sylvania
- Philips Lighting
- Sunpack Electronics
- Vadco

Consumer Reports ranked the Osram Sylvania Model CF30EL/C/830/MED/6 as its top choice for a CFL to replace a 100-watt incandescent lamp. But it found all the Osram Sylvania, GE, and Philips lamps it tested to be satisfactory. They all met their claims for the amount of light produced and the lifetime of the lamp. In fact, *Consumer Reports* found that some GE, Panasonic, and Sylvania lamps had median lives more than 80 percent longer than their 10,000 hour claim.

Most CFLs cannot be dimmed without doing damage to the lamp. Be careful not to dim any CFL that does not explicitly state on the package that it is dimmable. The only dimmable model I'm aware of in the U.S is the Philips Earth Lite.

Early CFLs with magnetic ballasts had a problem with flicker, but that has disappeared among newer lamps. All fluorescents produce a bluish light that some people find cold and harsh, but some CFLs are treated with phosphors that convert it to a warmer light. For all purposes that require a lamp to be lit for long periods of time, CFLs should be seriously considered.

Halogen Torchieres

A few years ago, some colleges that had retrofitted their classrooms with very efficient lighting were startled to see their electricity bills go *up*. They discovered that droves of students had brought halogen torchieres into their dormitory rooms. About 50 *million* of these lamps now are in use across the country.

The halogen torchiere is the lighting equivalent of the sport utility vehicle. It casts an extremely bright light straight upward, where it reflects off the ceiling and fills the room. Halogen torchieres suck up 300 to 500 watts of power, and can add $100 or more to your yearly electricity bill.

Halogen torchieres are also dangerous. A 500-watt lamp gets so hot (1,230°F) that you can fry an egg on a pan placed on top of the fixture. It can easily ignite curtains or other flammable material. Some colleges (wor-

---

T O O L B O X

All fluorescent lamps contain a small amount of mercury. As CFLs have become more common, many communities have set up recycling programs for them, in which the mercury is recovered and used again. If your community has no such program, agitate for one. Until then, dispose of CFLs where you drop off other hazardous household wastes (e.g., batteries, solvents, and paints).

ried about fire liability) will pay students the full cost of replacing their halogen torchieres with more expensive compact fluorescent torchieres that draw only 50 to 80 watts and are cool enough to touch.

High-Intensity Discharge Lamps

For rooms requiring a great deal of light, one manufacturer, Microsun, now makes a pricey but effective metal halide high-intensity discharge lamp. This type of lamp, typically used as street lighting, produces more light per watt than any other common lamp. Microsun's lamp renders color accurately (unlike most streetlights). I bought one as a desk lamp while preparing this book, and it works well. It produces more light and warmer light than my compact fluorescent torchiere.

Independent lab tests showed that a 68-watt Microsun lamp delivers a whopping 6,000 lumens—more light than a 300-watt halogen torchiere. The Microsun bulb lasts 10,000 hours and costs about $25. Unfortunately, it requires a particular fixture available only from Microsun. Call toll-free 888-526-0033 for a catalog. These fixtures, in the form of attractive floor and table lamps, are expensive. The whole package, fixture plus lamp, starts at $245.

E N E R G Y   E X T R A S

It is estimated that about three light bulbs burn five or more hours a day in the average U.S. home. If those three bulbs were replaced with compact fluorescents in every home, it would eliminate about 23 million tons of CO_2 emissions a year! If every household replaced just the bulbs it uses most frequently with CFLs, the United States would reduce its yearly electricity use by 32 billion kilowatt-hours.

Daylighting

The best lighting source, when it is available, is the sun. It is free, requires no electricity, and does not pollute. The sun provides light in a visible spectrum to which the human eye has grown accustomed over hundreds of thousands of years. In office buildings designed to draw natural daylight into their interiors, workers are happier, more productive, and take fewer sick days. In hospitals, patients with a window recover faster.

With spectrally selective windows now available, daylighting need not conflict with comfort. Clerestories, skylights, light shelves, and light pipes are innovative ways to bring sunlight into interiors. There is an art to designing and installing such devices, however, so consult with an architect or other lighting expert when considering alterations or additions to your home. 🌎

Chill: Refrigerators

Your refrigerator may account for up to 25 percent of your electricity bill!

What to Look For

The typical refrigerator sold in 1973 used about 2,000 kilowatt-hours per year. At 8.3 cents per kilowatt-hour (the average electricity rate today in the U.S.), you would pay $166 a year to run it. Since the average refrigerator has a life of 14 years, its lifetime electric bill would be $2,324—at least three times the initial purchase price of the appliance.

The *most efficient* new refrigerators use about a fourth as much electricity, or 650 kilowatt-hours per year. In fact, today's *average* refrigerator uses less than a third as much electricity as the average refrigerator in 1973, although it (1) is larger; (2) has new features, like automatic defrost and ice through-the-door; and (3) uses substitutes for the ozone-destroying CFCs that used to flow in its chilly veins.

Moreover, by July 1, 2001, federal regulations say that the average new refrigerator must use 30 percent less than it does today! The average refrigerator will then use only as much electricity as a single 40-watt bulb left on all the time, and its lifetime electrical bill will be about $400.

Much of the heat gain in super-efficient refrigerators occurs via conduction through the sides and top. The smaller the refrigerator, the greater the amount of external surface there is per unit of internal volume. So the energy penalty for a larger refrigerator is not as great as with a larger-than-needed furnace or lamp, and can be compensated for with more efficient features. For the same reason, it is always more efficient to use one large refrigerator than two small ones.

All refrigerators are vastly more efficient than they were ten years ago, but there are still important differences among them. For a list of the most efficient models of all standard sizes (20 percent more efficient than required by law), see the ACEEE Web site (*http://aceee.org/consumerguide*) or the *Consumer Guide to Home Energy Savings, Seventh Edition* (Washington, D.C.: American Council for an Energy-Efficient Economy, 1999).

As a general rule, the smaller an appliance, the less energy it uses. But note that Jenn-Air's 18.5 cubic-foot model uses less energy than any smaller refrigerator, and Jenn-Air's best 21 cubic-foot model uses the same amount of electricity as the best 15 cubic-foot models.

How to Use Them

• Check the settings: the fridge should be between 38 and 42 degrees Fahrenheit and the freezer should be between 0 and 5 degrees. If your settings are only 10 degrees too cold, your energy bill will be 25 percent higher than it needs to be.

• Check the seals for cracks and dried-on food.

• Keep the condenser coils clean. (Sears sells a special long-handled brush for this purpose.)

• Food retains cold better than air does, so a full (but not over-crowded) fridge is more efficient than an empty one.

• Don't locate your stove right next to your refrigerator. Heat from cooking will increase the amount of work the fridge must do.

Refrigerators are the bad boys of energy use, so buy a fridge that is in line with your actual needs. 🌍

ENERGY EXTRAS

If you are turning to solar cells to power your house, or your kitchen, you should seriously consider a Sun Frost refrigerator. P.O. Box 1101-W, Arcata, CA, 95521-1101. Tel. 707-822-9095. Fax: 777-822-6213, http://www.sunfrost.com. *The Sun Frost is a super-efficient refrigerator that comes in a wide variety of sizes and shapes, and can be configured to run directly off DC current from solar-charged batteries. It can also be set up to run off regular alternating current in anyone's home. Ten years ago, the Sun Frost was more efficient than any competing refrigerator, but, under government pressure, the industry has erased most of the gap.*

Ways to Cook a Planet: Stoves and Ovens

Microwave ovens use only a third as much energy as regular ovens.

Stoves: What to Look For

Stoves are major energy users. Natural gas stoves are much more efficient than electric stoves because of conversion losses in generating electricity and transmitting it over power lines. Typically, three to four units of fuel are consumed for every unit of heat delivered to an electric stove. With natural gas, on the other hand, the fuel is burned right under your pan and the heat-losses are avoided.

There are some creative electric stoves. Halogen stoves use

a quartz-halogen lamp to radiate heat to the ceramic glass surface. Magnetic induction stoves heat iron or steel pans directly by exciting the molecules magnetically. Both are more efficient than standard electric coil stoves, but they are very expensive and the energy savings can't justify the price.

How to Use Them

• Cover pots to stop heat from escaping. This trivial step can reduce energy use in cooking by up to two-thirds. A pressure cooker saves even more energy.

• Use the smallest pan (and on electric stoves, the smallest burner) that will do the job.

• With electric stoves, use only flat-bottomed pans that make full contact with the stove elements. A rounded or warped pan can waste more than half the energy.

Ovens: What to Look For

Conventional ovens are inherently inefficient. Typically, only 6 percent of the energy an oven uses actually heats *food*. The rest heats up about 35 pounds of steel and a lot of air inside the oven. Since there are no federal efficiency labels for ovens, the best guide to energy efficiency is to buy an oven with a self-cleaning function. These are better insulated, and they therefore use 20 percent less energy than other ovens.

Convection ovens are more efficient than conventional ovens, especially if you put multiple dishes in the oven at the same time. In a conventional oven, the air remains still. In a convection oven the air is circulated, to assure that the heat it carries is constantly being delivered to all the surface areas— food and containers—inside.

How to Use Them

• Cook more than one dish in the oven at the same time.

• Roast or bake large portions (e.g. a whole turkey) and then re-heat the excess later using a more efficient stove

---

ENERGY EXTRAS

By eating less meat you save energy. It takes seven pounds of grain to produce one pound of beef. It takes four pounds of grain to get a single pound of pork, two pounds to produce a pound of poultry. If Americans trimmed their annual meat consumption to the level of Italians, we would save 105 million tons of grain— enough to feed two-thirds of India for a year. Besides, you always have to refrigerate and then cook meat before you eat it, using more energy. Eating lower on the food chain not only saves energy and reduces global warming, but it also makes nutritional sense and saves money.

---

or microwave, so that you don't have to heat up a big oven more than once.

• Use thermometers to let you know how the food is progressing, so you don't waste energy by overcooking.

• If you use the self-cleaning function, use it immediately after you've baked something, so that you don't have to spend energy heating up a cold oven.

• Turn off the oven a few minutes before the food is finished and let the residual heat complete the job.

Microwaves

Microwave ovens heat food by exciting water and fat molecules inside the food itself; they don't waste energy heating metal and air. Most new microwaves have intelligent sensors that turn off the oven when the food is finished. Microwaves do not work for all foods, but they are super-efficient for the large number of foods for which they are suitable. Check your instruction manual to learn how to use your microwave safely. 🌍

Clean Plates: Dishwashers

Over the long haul of life on this planet, it is the ecologists,
and not the bookkeepers of business, who are the ultimate
accountants.—Stewart L. Udall

What to Look For

Most of the energy dishwashers use is to heat the water. The most efficient models use less than half as much water as the least efficient ones. You'll be happy to learn that it's also more efficient to use a dishwasher than to wash dishes by hand. A water-saving aerator nozzle on a kitchen sink typically has a flow of 2.5 gallons *per minute*, whereas the *total* water consumption for a full load in an efficient dishwasher is less than five gallons. Efficient dishwashers also have thermostats that guarantee that they use only as much energy as necessary to heat their water to 140° F. (Older models had their heating elements on timers, and pumped in several minutes of electricity whether it was needed or not.)

A few high-efficiency European models are available in specialty shops in America, of which the super-efficient Asko is the most widely available. Some domestic brands have begun responding with models that far surpass federal appliance efficiency requirements.

Dishwashers come in various sizes and configurations. In addition to energy savings, it is important to choose a model that allows a large number of dishes to be included in a single load. Moreover, some dishwashers do a much better job than others of cleaning dishes! Once you have your list of acceptably efficient dishwashers, check *Consumer Reports* magazine for an evaluation of these other features. (All your energy savings disappear if you have to pre-wash your dishes, or run them through two complete cycles, to get them clean.) For a list of dishwashers that exceed the 1994 energy efficiency standard by at least 30 percent, see the ACEEE Web site *(http://aceee. org/consumerguide)* or the *Consumer Guide to Home Energy Savings, Seventh Edition* (Washington, D.C.: American Council for an Energy-Efficient Economy, 1999).

How to Use Them

• Wash only full loads.

• You'll get more pieces in the washer if you wash the big stuff by hand.

• Use the air-dry setting or turn the dishwasher off after the final rinse and open the door.

• Use short cycles.

• Use the booster heater so that you can turn your hot water tank to a lower temperature.

Some people worry that if their water heater is not at its highest temperature, their dishwasher won't kill germs. But the highest setting on both the water heater and the dishwasher is 140° F— far too low to disinfect anything. Dishes are cleaned by a combination of detergent and friction from agitated warm water. Dishwashing isn't designed to kill germs; it just washes them down the drain. To kill most germs, you would have to raise the water temperature at least to its boiling point (212° F). No dishwasher does that. 🐾

Clean Clothes: Laundry

A typical American household does four hundred loads of laundry a year, which use about 16,000 gallons of water.

Washing Machines

Even if you use warm water instead of hot, heating the water will account for 85 percent of the total energy used to wash your clothes. The average U.S. washing machine uses four times as much water as the typical European washing machine. The difference is that most European clothes washers operate on a horizontal axis (like a Ferris wheel) while, until recently, all American models operated on a vertical axis (like a merry-go-round).

Vertical-axis washing machines always use more water. A

vertical-axis machine must be filled with water to the point where all the clothes are covered, and then the agitator swishes them back and forth. In a horizontal axis machine, the tub itself rotates, tumbling the clothes through a much smaller amount of water. Horizontal axis machines save water, electricity, detergent, fabric wear, and drying costs (because they extract more water in the spin).

In the mid-1990s, a handful of European manufacturers (AEG, Miele, and Asko) began to invade the American market with horizontal-axis washing machines. They were substantially more expensive than standard washing machines, and became something of a status symbol among certain environmentally-aware yuppies.

As federal efficiency standards tightened, more and more American manufacturers began producing efficient horizontal-axis machines. While they are still more expensive than vertical-axis machines, competition is bringing the price down. Even at current prices, they make economic sense if you use the warm setting for most loads.

The warm setting is sufficient for virtually all washing needs except diapers or oil stains (which will probably require an enzymatic detergent). A warm pre-soak may be necessary for very dirty clothes. I have gone years without using the hot setting.

Clothes Dryers

A clothes dryer usually uses more energy than any other appliance except the refrigerator. A comparison of clothes dryers to refrigerators demonstrates the importance of federal appliance efficiency standards. Refrigerators are covered by the program; dryers are not. The average refrigerator today keeps your food just as cold while using only *one-third* the electricity it used 20 years ago—even though a ban on CFCs (to protect the stratospheric ozone layer) forced refrigerator manufacturers to turn to less efficient refrigerants. Federal standards helped produce a technological revolution that benefits everyone.

Clothes dryers, on the other hand, have seen no meaningful progress in the last quarter century. As a consequence, my 17-

year-old dryer is as sophisticated and efficient as any new machine. The major manufacturers have not even experimented with the most obvious prospective energy-saving technologies (such as microwave dryers).

All U.S. dryers use the same basic technology. As with stoves and furnaces and anything else producing heat, gas is inherently more efficient than electricity. The next biggest variable is the "termination control." This morbid, CIA-sounding phrase refers to the device that turns the dryer off.

The most energy-efficient control is also the most sophisti-  cated: a moisture sensor that tells when the clothes are dry. Moisture sensors are tried, true, and cheap. Buying a dryer that uses a temperature control or a simple timer is a decision to throw away money on energy, and to risk overdrying (and prematurely aging) your clothes.

The most intelligent choice of dryer is, of course, the clothesline. Precisely the same amount of energy is used to evaporate water from clothes on a line as in a dryer. The difference is that solar energy is free and renewable, and it doesn't contribute to global warming. 🌏

Power Shift

The average American creates 4.5 tons of carbon dioxide from home electricity use every year.

Most consumers believe their electricity comes from cleaner sources than it actually does. When Colorado, for example, passed a new law that makes utilities disclose the source of their electricity, customers were amazed to learn that 93 percent of their power comes from coal. Electric utilities release more greenhouse gases and other air pollutants than any other industry in the United States.

Historically, environmental groups have lobbied regulators to require utilities to install renewable energy plants, or at least to buy green power from independent green-power producers. This has led to spectacular payoffs in states such as Minnesota, where regulators have required large-scale investments in wind farms.

In states that allow utilities to compete for customers, consumers can choose a less polluting source of electricity for themselves. In California, for example, there are already six retail, and five wholesale, power suppliers offering environmentally-superior products.

The environmental movement also has taken the lead in certifying that green power is genuinely superior. This is a tricky job. Determining the source, and verifying the cleanliness, of electricity is an impossible task for an individual consumer. The electrons you buy are invisible, and they arrive on the same power lines as your neighbor's electrons. You literally pay for something you never see. If you choose to pay a premium because there is a wind turbine operating somewhere out there on the grid instead of a coal plant, it is wise to have your power source independently certified by an objective third party.

The "Green-e" mark of the Center for Resource Solutions is currently the best seal of environmental approval for electricity. To get more information about Green-e, visit its Web site at *http://www.green-e.org*.

When you see the Green-e logo on an advertisement for electricity, you know that:

- At least 50 percent of the electricity supply for the product comes from renewable resources.

- One year after deregulation, the product must contain at least 5 percent renewable electricity. Moreover, this power must come from a new source—to encourage the development of additional green-energy facilities. This way, the utility can't just charge you more for electricity from wind turbines that were required by regulators in years past. This renewable-source requirement increases to 10 percent in the second year. Green-e intends to increase the new renewable requirements 5 percent each year until 25 percent of the total product content is from new renewable resources.

• Any nonrenewable part of the product must have lower air emissions (including carbon dioxide) than your traditional mix of electricity would have had if you hadn't switched.

• The company offering the product must agree to conduct an annual third-party audit to ensure that they have purchased or produced enough renewable power to cover what they have sold to customers.

Toyota's California production facilities, Patagonia, and hundreds of Episcopal churches are among those who have signed up for Green-e certified power.

Green power is not the ultimate solution, because it relies upon people to pay more than their neighbors for identical electrons. In return, they get only the secret satisfaction of knowing that they are creating demand for clean energy—and helping to save the planet. Many people will do this, motivated by the same sentiments that causes them to send money to the Sierra Club. But it is an act rooted in environmental values, not economic motives.

Paying a premium for green power is exactly the opposite of what would prevail in an ideal world. Energy sources that impose no global warming, acid rain, smog, or other burdens on the public and the environment should be less expensive than those that do. If someone wants to use coal as his source of electricity, society should force him to pay through the nose.

Today, however, green power is the best way we have to assure that the renewable-energy industry does not fail once again, no matter what the political climate. An extra few bucks a month is a small price to keep the world's most promising energy sources viable until reasonable prices usher in a solar revolution. 🌍

The Three Green Rs

Listen up, you couch potatoes: each recycled beer can saves enough electricity to run a television set for three hours.

For the past decade, most schoolchildren in America have been exposed to the reduce, reuse, recycle hierarchy. It makes sense from every environmental perspective, including climate protection.

The relationship between raw materials and global warming is straightforward. We mine raw material such as ores; transport them, refine them, turn them into consumer goods, use the stuff (often only briefly), and then haul it to the dump. Each step of the process requires energy, sometimes prodigious amounts of energy. Another raw material, trees, would in fact be useful if left in place to absorb CO_2 and protect hillsides from torrential downpours. The less waste we produce, the less energy we use, and the less carbon dioxide we generate.

Reduce

While many of us think that spending money to keep up with (or ahead of) the Joneses has outlived its utility as a status symbol, *reduce* is not a call for pain and suffering, freezing

---

ENERGY EXTRAS

In the first year after the California utility industry opened to competition, about 1 percent of all consumers in California chose to buy electricity from a green source. That is only half as long as it took new telecommunications companies to obtain 1-percent market penetration after they were opened to competition. This is especially impressive when you consider that the telecom companies were offering cheaper service, whereas the green power companies are charging a premium for their power.

---

in the dark, or romanticizing the stone age. In fact, it is just the opposite.

A principal source of waste is junk, goods that are made with no reference to quality and durability. Once, a piece of furniture was cared for and passed on to future generations. As time passed and the piece aged, it became even more cherished and valuable. Today, a piece of furniture is more likely bought on a whim, used casually for a few years, and eventually hauled to the landfill. Chances are its workmanship is shoddy, the materials poor, and its design soon outmoded.

To reduce means to buy durable products with longer useful lives instead of throw-away products. It means to take more care in what you buy. Consider goods as investments that accumulate as part of your wealth, instead of as consumption that depletes it. Assess what you really need, eliminate the clutter, and find a way to assert your individuality and status other than through the volume of your possessions.

Reuse

Four out of five glass containers that are manufactured today are used just once and thrown away. In comparison, my mother used the same glass jars for canning over and over. Twenty years after I graduated from college, she was still using jars she'd used before I was born. (When she died, those jars were in her cellar, filled with fruits and vegetables she'd put up herself.) If you compare the energy efficiency of my mother's use of glass to the energy efficiency of any glass-using business today, she would be off the top of the charts. But my mother was in no way unusual. Her behavior was the norm for that era.

It often makes both environmental and economic sense to *reuse* things repeatedly, and even to buy used goods. I printed out the draft manuscript of this book on an ink jet printer that, in the parlance of the trade, was pre-owned. Someone else once used this printer and decided, for whatever reason, that he wanted a new printer. A company took this one in trade, refurbished it, and sold it to me for $195 with a six-month warranty. It is small, lightweight, and fits neatly in my computer case. It suits my needs and my budget. My pre-owned Honda Civic was a steal, because a car's value drops sharply the moment it is driven off the lot. The brass bed we found on a Virginia farm 30 years ago is as classy and useful as it was a century earlier. To reuse is not necessarily to sacrifice. (We even have a pre-owned animal companion—formerly known as a pet.)

Some European countries, most notably Germany, Sweden, and the Netherlands, are experimenting with the concept of having a company take lifetime custody for the materials inside its products. Most companies are kicking and screaming about this, but the impact for the consumer is beneficial. If BMW knows that it is responsible for all the materials in its cars when they are disposed of, it will take far greater care in designing cars that can be repaired, updated, reused, and recycled. Such an approach might help to reverse the energy-wasting, throw-away culture that has come to dominate much of American industry.

Recycle

Forgive me if I crow a little here. Largely as a result of Earth Day's long-term emphasis on it, recycling as a part of everyday life has become commonplace all over America.

Not that we didn't have a rocky start. Three thousand recycling centers were organized around Earth Day 1970; many failed, and others endured only as marginal operations. It takes a great deal of commitment to sort and save everything recyclable and to haul it periodically to a distant center. Since most of this hauling was done by car, it was also probably hauled at a net energy loss.

Recycling has since become much easier—and much more energy efficient—with widespread curbside recycling. In 1988, when we first began planning Earth Day 1990, there were fewer than 1,000 curbside recycling programs in America. The Earth Day staff drafted model curbside recycling ordinances and distributed them to every mayor and city council, along with a note saying, essentially, "This would be a terrific thing to pass and announce on Earth Day 1990." We encouraged our organizers across the land to press for recycling initiatives. Today there are about 9,000 curbside recycling programs serving 135 million people (about half of the nation's residents).

ENERGY EXTRAS

Aluminum-can manufacturers have been making cans lighter, stretching each pound of metal to make more cans. In 1972, each pound of aluminum produced 22 cans; today, it yields 29 cans.

A recycled can is first cleaned, delacquered, and shredded. The metal is melted at 1,400°F, and various salts are added. When the molten metal has the right chemistry, it is poured into large rectangular ingots that can weigh up to 20 tons. Later, the cooled ingot is passed between two giant steel rollers in a rolling mill until it is a sheet about a half-inch thick and a thousand feet long. This sheet is then taken to a finishing mill where it is annealed to soften it, and passed through a series of rollers until it is paper thin. The finished coil—which may be two miles long and contain 1,200,000 recycled cans—is rolled up for shipment to a can manufacturer. Then the whole cycle begins again.

Recycling can save staggering amounts of energy. On average, this is how much energy is saved by recycling the folowing materials versus obtaining them from virgin resources:

Aluminum	95 percent
Steel	61 percent
Newsprint	45 percent
Typing Paper	35 percent
Glass	31 percent

Imagine how much more energy we could save by increasing our recycling rates for these materials. For now, we recycle just 38 percent of aluminum, 35 percent of paper, 32 percent of steel, and 23 percent of glass.

The three types of plastic that are most commonly recycled also offer large *potential* energy savings. But these savings are mostly hypothetical, since less than 5 percent of plastics are recycled.

PET (polyethylene terephtalate)	57 percent
HDPE (high-density polyethylene)	75 percent
PP (polypropylene)	74 percent

Plastic is bulky, it quickly fills up trucks, and many products combine multiple types of plastics that cannot be recycled together. These are not intractable problems. But unless manufacturers are given some responsibility for the lifetime custody of their products (so that ease-of-recycling becomes a design criterion), the potential energy savings from recycled plastic will likely remain unrealized.

Complete the Circle

When you place your recyclable materials at the curb, you are not really recycling. You are merely sorting. For something to be recycled, it must be made into a new product and reused. Unless there is a solid predictable market for recycled materials, private firms will not invest in the facilities to recycle, and market prices for recycled commodities will fluctuate wildly.

It will be wonderful when recycling is so routine that consumers won't need to check the recycled content of a product because *everything* will be made mostly of recycled materials. That is already true for some paper, for example, and largely true for aluminum cans. (Sixty-three percent of aluminum cans are now recycled.) But for too many other goods, it remains rare.

For now, we as consumers must look for products that have the highest percentage of post-consumer recycled content. By doing so, we can establish a market, send a message to manufacturers, and close the circle.

T O O L B O X

An interesting experiment is underway at Interface, Inc., a Fortune 500 carpet manufacturer whose chairman, Ray Anderson, may be the "greenest" industrialist on Earth. Under no pressure from anyone, Anderson developed a carpet that can be 100 percent recycled and composted. Interface, which mostly sells to large commercial customers, leases this carpet under what it terms its "Evergreen Lease." When the Evergreen carpet is worn, Interface will replace it with new carpet. The old carpet is then completely recycled. The energy savings are estimated to exceed 95 percent. Interface will also recycle your old commercial carpet, whether you purchased it from Interface or not. If you are in a position to influence carpeting decisions within your company, urge it to switch to an Evergreen Lease from Interface. The best way to contact the company is through their Web address www.interfaceinc.com. *If you don't have computer access, it's 2859 Pares Ferry Road Suite 2000, Atlanta, GA 30339, 770-437-6800.*

Troubleshooting: Clean Power

Making changes in your personal
use of energy is an important first step.
But personal choices are not a
substitute for political action.
Global problems can only be solved
through public policies that engage
everyone in the shared enterprise.
To change public policies governing
energy, we need to mobilize for a
political struggle against the world's most
powerful economic interests.

If the first Earth Day had not led directly to the creation of the EPA and the passage of the Clean Air Act, the environmental movement would not have acquired the momentum it attained. Similarly, this millennial Earth Day needs to have political consequences. The objective of Earth Day—and this book—is to engage tens of millions of citizens to demand policies that produce a sustainable future for themselves and their children. This section explores the kinds of basic agendas against which we should be measuring progress, as well as specific policies that will help us reach our common goal: the safe keeping of our planet.

America, the Skeptical

Oil is much too valuable to be burned to send sports utility vehicles to the mall. Our great-grandchildren (and all subsequent generations forever) will curse our self-indulgence as they spend scarce, expensive resources trying to reverse our waste and synthesize these complex molecules from carbon and hydrogen.

Many hugely powerful oil, coal, and electric corporations genuinely fear a world in which the use of fossil fuels declines. These firms see their own fortunes as being dependent on burning more carbon each year. They have fought hard and successfully to thwart laws, regulations, treaties, and policies that would reduce the use of fossil fuels.

Smoke and Mirrors

The propaganda put forth by these behemoths mirrors the cigarette industry's denial of a link between smoking and lung cancer decades after virtually *everyone* had stopped believing the lies. A recent $5-million, tobacco-style PR initiative by the

American Petroleum Institute, Exxon, Chevron, and the Southern Company, among others, was scuttled with embarrassment after an anonymous insider leaked the planning documents to *The New York Times*. Among other things, the initiative would have funded a phony propaganda machine called the "Global Climate Science Data Center," and sought to persuade the media, members of Congress, governors, state legislators, and the public that this Center was an objective, independent source of climate information. To hide its connections to the fossil-fuel industry, funding for the plan would have been funneled through front groups with misleading names such as the "Committee for a Constructive Tomorrow" and the "Frontiers of Freedom."

The carbon fuel industry's most brazen argument is that America can't afford to lead a transition to solar energy. Its ads proclaim that even modest reductions in America's CO_2 emissions—the tiny, measured steps agreed to in Kyoto—will lead to economic collapse.

This public-relations blitz demands an answer. The most appropriate response is a good horse laugh. Let's look at the facts.

In terms of wealth, the United States' status is without precedent. Americans own a stunning $39 trillion in financial assets. The U.S. stock market now accounts for more than half the value of all the equity markets in the world.

Just since 1990, America's gross domestic product has increased from $5.6 trillion per year to $8.7 trillion. If those numbers are meaningless to you, think of it this way: just the last nine years' *growth* in the U.S. economy is greater than the *total* combined value of all the goods and services produced by the world's 113 poorest countries.

In 1998, the 4.5 percent of the world's people lucky enough to be Americans were responsible for almost half of the entire world's growth in the demand for goods and services. (Part of that demand is from the 11 million Americans who are millionaires, and the 200 who are billionaires.)

In contrast, according to World Bank figures, 81 percent of the world's population would qualify for food stamps if they lived in the United States. The 50 nations of sub-Saharan Africa (minus South Africa)—with a combined population more than twice that of the United States—have a *combined* gross domestic product of $198 billion, roughly that of the State of Virginia.

We can't afford to lead the world? If not us, who?

Oil and American Wealth

For a century after the American Revolution, America was an economic subsidiary of the British Empire. Our industries, our railroads, and our cities were built with British capital. Great Britain, then the world's richest country, dominated the age of coal.

In the closing years of the nineteenth century, America began to lead the world into the age of oil. For the next half-century, we were the world's foremost oil producer. Even today, 30 years after U.S. oil production peaked and began to decline, American oil companies still dominate the world petroleum industry, and American technology still produces and refines the world's oil.

Many factors went into America's transition from a developing country of great promise 125 years ago to the world colossus it is today. These included universal public education, democratic government, bountiful natural resources, efficient capital markets, an entrepreneurial culture, social mobility, official protection for intellectual property, and the great good fortune of having had two world wars fought elsewhere using ships, planes, and tanks manufactured here.

To top it off, the easily-tapped oil fields of Pennsylvania, Texas, Louisiana, and California became available just in time to power America's development into a modern industrial state.

The United States has been the world's wealthiest country throughout the lives of everyone reading this book. Most Americans have come to think of this wealth as our due. But things can change with time. If the European Union evolves into a genuine United States of Europe, it would have a larger combined economy than ours (and possibly a higher per-capita GDP, as well). China has vast resources and a huge, disciplined population committed to economic growth. Russia could come back from the brink.

T O O L B O X

It is astonishing how well renewable-energy technologies have done despite this unremittingly hostile political environment. In Winner, Loser or Innocent Victim: Has Renewable Energy Performed as Expected?, *four distinguished energy economists from Resources for the Future conclude that the costs of renewable-energy technologies have fallen about as fast as early proponents predicted. "This is remarkable," they write, "given that renewable technologies have not significantly penetrated the market, nor . . . (achieved) economies of scale in production, as many analysts anticipated when forming their cost projections."*

The RFF team notes that federal spending on research, development, and demonstration for carbon fuels was almost ten times greater than that for renewables during the Bush Administration. And even that, of course, pales when compared to the tax loopholes, below-cost leases, military subsidies, and other government aids to keep the oil flowing. If renewable energy sources had enjoyed support comparable to the rich subsidies and other tangible benefits enjoyed by carbon fuels over the last two decades, the solar-energy revolution now would be comparable in scope and impact to the information revolution. Winner, Loser or Innocent Victim *is available on the Internet at* www.repp.org.

Continued American leadership will almost certainly depend upon continued technological leadership. And in a greenhouse world, the nation leading the commercialization of the energy technologies that safeguard the climate—solar cells, fuel cells, hydrogen generators, wind turbines, biofuels, geothermal power plants, hypercars, and the rest—will have a competitive advantage comparable to that provided by cheap oil a century ago.

Twenty-five years ago, America led the world in most renewable-energy technologies. Venture capital was flowing to wind farms and solar manufacturers. Our National Renewable Energy Laboratory in Golden, Colorado, was at the cutting edge of disciplines, from photoelectrochemistry to basic-materials science. Then many of the most dynamic young solar companies were bought by oil companies, which dismantled them after the Reagan election. The Reagan Administration slashed NREL's budget by more than 80 percent and swiftly abandoned the Carter Administration's goal of getting 20 percent of our energy from the sun by the year 2000.

The U.S. decision to abandon solar energy was unambiguous, and the forces behind it were not mysterious. The oil industry and its political pals shut it down. I was in a position to observe closely this exercise in raw political muscle. It has cost us a quarter-century of progress.

America's national interest

> **ENERGY EXTRAS**
>
> *We first turned to "rock oil" or petroleum in the middle of the nineteenth century as a result of an energy crisis. The Civil War interrupted the supply of whale oil for lamps, and petroleum (from the Greek words for rock and oil) was developed as a substitute. Petroleum was soon found to have characteristics that made it superior to coal for myriad energy applications, and with the advent of the gasoline-powered automobile, the oil industry boomed.*

is diametrically opposed to the narrow interests of the carbon-fuel industries. This is not because oil and coal companies are run by evil people. The explanation is less dramatic than that.

For mundane organizational reasons, industries rarely make a successful transition to the newer technologies that displace them. Manufacturers of fountain pens did not make the transition to typewriters. Manufacturers of carbon paper did not transform themselves into manufacturers of photocopy machines. Radio manufacturers failed to dominate the television market. Manufacturers of railroad engines did not begin making trucks or airplanes.

Most CEOs of carbon-fuel companies contend that global warming is a myth, all the evidence to the contrary notwithstanding. Global warming means they have spent their entire careers rushing down a blind alley. Psychologically, their sense of self-esteem requires complete denial. Moreover, they have risen to the top of a macho corporate culture that offers its richest rewards to those who successfully wildcat new oil fields while overcoming physical and political obstacles in remote corners of the world. Naturally, they laugh at the idea of running a modern industrial state on sunshine. If these men continue to dominate American energy policy, we will continue down the path to disaster.

America has the wealth, scientific excellence, engineering skills, educated work force, decentralized decision-making, democratic institutions, and supportive public necessary to lead the world into a solar-powered era. If we fail to do so, we will eventually be led there by someone else. 🌐

Programs, Not Promises

Just the increase *in U.S. emissions during the last decade, all by itself, would rank as the seventh largest source of CO_2 in the world.*

Much of the debate in the United States on global warming has consisted of arguments over distant CO_2 emissions goals. In an

ENERGY EXTRAS

In the 1970s, a huge amount of effort went into convincing President Carter to embrace a goal of obtaining 20 percent of the nation's energy from renewable sources by the year 2000. After exhaustive studies, including a detailed analysis produced by a consortium of several national laboratories and universities, Carter adopted the goal in the last few months of his Administration.

Within six months of President Reagan's inauguration, the new Administration had shredded the solar commercialization program, decimated the federal research and development effort, and even yanked the solar collectors off the White House roof. In retrospect, Carter's interminable, Hamlet-like deliberations over the "correct" goal seem silly. They were utterly and completely irrelevant in the post-Carter years.

electoral system where sea changes occur every two and four years, these distant goals are like an ever-receding shore. In 1993, President Clinton, under prodding by Vice President Gore and environmental leaders, pledged to reduce carbon-dioxide emissions to 1990 levels by the year 2000. But there was no strategy, no regulatory initiative, and no budget to do anything to achieve that goal. Naturally, nothing happened. As a consequence, the nation's CO_2 emissions in 2000 will be 12 percent higher than in 1990.

Twelve percent doesn't sound like a lot, but it is 12 percent of a huge figure. The 12 percent *growth* in U.S. emissions over the last nine years is more than the *total* CO_2 produced this year by England or France. It is *far* more than is produced by South America—Brazil, Argentina, Columbia, Chile, Ecuador,

Bolivia, Uruguay, Paraguay, Panama, Guatemala, Honduras, and Nicaragua—combined.

In 1998, President Clinton announced that his *new* goal was to reduce America's CO_2 emissions to the same level as 1990 sometime between 2008 and 2012 (i.e., four years after the end of the *next* president's *second* term). Obviously, future presidents will have their own goals, programs, and agendas, and will not care much about President Clinton's.

So why does the Earth Day Clean Energy Agenda (see page 162) have long-range numerical goals? These are not carved-in-stone tablets, brought down from Mount Sinai, but rather indicators that give us a sense of direction and help us gauge whether our intermediate steps are ambitious enough. These long-term targets were adopted by a broad cross-section of top energy experts according to what could be achieved if we get *serious*. Meeting them would not involve a war-time mobilization, but it *would* involve a level of commitment comparable to, say, the Interstate Highway Act.

Over the next couple of decades, the United States is in a position to get as much safe, sustainable energy as we Americans are willing to pay for. What *counts* are not the abstract goals spewed forth by our leaders but the laws that Congress actually passes, and the programs that future presidents actually implement. Goals provide inspiration, but dollars build wind turbines and solar-power plants. Dollars can insulate dwellings and develop fuel-cell-powered automobiles. Dollars can replace natural-gas pipelines with hydrogen pipelines. Dollars can clean up the air we breathe.

When politicians proclaim goals that seem to promise you the moon, reply with a bit of folk wisdom made popular by the movie *Jerry McGuire:* "Show me the money." 🌎

Ban Coal

Both China and the U.K. are offering incentives to solar-cell manufacturers to locate near abandoned coal mines.

In addition to producing more CO_2 than any other energy source, coal also releases particulates and sulfur dioxide, which cause half a million premature deaths each year and afflict millions of other people around the world with respiratory diseases. Coal smoke from cooking in rural areas may account for an additional 1.8 million annual deaths. Acid rain from coal has laid waste to vast tracts of forests in Europe, Asia, and North America. New mining techniques remove whole mountain tops and dump the tailings in adjacent valleys, leaving a moonscape behind.

China is the world's largest coal producer. It mined 50 percent more coal last year than the United States, and it depends on coal for 73 percent of all its energy (not just its electricity, but all its energy). Yet China has cut its coal subsidies in half since 1984, and it will have shut down 14 large state-owned coal mines and 25,000 small coal mines by the end of the year 2000. Beijing has established 40 coal-free zones, and (along with Shanghai, Lanzhou, Xian, and Shenyang) ultimately plans to ban coal from the city. Japan, Britain, France, Belgium, and Spain all have cut their coal use in half in the last 15 years.

Meanwhile, the United States still depends on coal for 53 percent of its electricity. If America is to take the lead in the transition to the solar era, these coal plants must be phased out over the next ten to 15 years. The first plants to go should be those that were grandfathered under the Clean Air Act and exempted from compliance. When it passed the Clean Air Act, Congress expected all these old plants to have been retired by now. Instead, their owners keep updating them and improving them with new boilers, generators, fuel-handling equipment, and whatnot—without installing pollution-control technology.

Employment transitions accompany any fundamental change. We've seen it in agriculture, horse farming, railroads, forestry, steel, tobacco, and many other fields. As a result of the United Kingdom's decision to scale-back its reliance on coal dramatically, the U.K. went from having 1,200,000 miners in 1978 to 13,000 today. China has eliminated 870,000 coal-mining jobs over the last five years, and has plans to retrain an additional 500,000 miners for other jobs in the next few years.

The United States, which employed 705,000 coal miners at its peak, now has only about 80,000. There are more Ph.D.s in West Virginia today than there are coal miners. In fact, there are more coal miners' widows in America than there are coal miners.

We will need programs to ensure a fair transition for affected workers. America should do everything it can to ease such transitions by providing new employment opportunities, training in those skills needed to exploit them, and financial assistance to bridge the gap. America has lots of valuable work for coal miners once they've stopped mining coal.

We must immediately stop turning to coal for fuel. If we burn the world's remaining coal, we will alter the global climate in ways that will diminish the wealth and productivity of human societies, as well as the biological diversity of the lands and the seas.

The best thing we can do with coal is leave it in the ground. 🌍

Here Comes the Sun

The environment is not something "out there."
We eat the environment. We drink the environment.
We breathe the environment. We are part of the environment,
and it is a part of us. We cannot destroy the environment
without destroying ourselves.

One of the most valuable things that government can do in the early stages of a product's development is to buy it in volume. In field after field, large-scale government procurement has brought down the cost of items as diverse as jet planes, drugs, and computers to the point where the private market could take

advantage of them.

The government should play a similar role in propelling the energy revolution. Large public-sector commitments to buy wind turbines, biofuels, fuel cells, hydrogen, hypercars, and other elements of a solar future will accelerate the speed at which such products become affordable for the rest of us. We typically think in terms of federal procurement, but state and local governments can play an important role too.

There is no more obvious a candidate for a federal buy-down than solar cells. Lowering the cost of solar cells would provide extraordinary public benefits. Solar cells make electricity, but they consume no fuel, produce no pollution, generate no radioactive waste, have long lifetimes, contain no moving parts, and require little maintenance. They can be fashioned mostly from silicon, which is the second most abundant element in the Earth's crust. Solar cells produce *zero* CO_2, the chief greenhouse gas.

Unfortunately, solar cells are not yet cheap enough to compete with heavily subsidized fossil fuels. Although the price of solar cells already has fallen about 40-fold, this technology remains roughly three times too expensive to achieve skyrocketing growth as a power source in the United States.

For a quarter-century, affordable solar cells have been the environmental brass ring, lying just outside the grasp of those who favor green power. Governmental procurement could lower their price to the point where they will take off on their own in the private sector. A comparison of the experiences of computer chips and solar cells vividly illustrates the value of government procurement in bringing new products to market.

Learning Curves

New technologies follow predictable learning curves. As production volumes increase, costs fall. For example, a product with an 80-percent learning curve will experience a 20 percent price drop when the volume is doubled. If production is doubled again, the price will decline an additional 20 percent.

This is common sense. If I gave you a pile of raw materials and said, *make me a car*, it would cost a fortune (and it probably would not be a very good car). If car-making grows into a cottage industry, prices will fall—perhaps into the range of a hand-built Ferrari. When volumes swell and you are manufacturing tens of thousands of cars per year, the marginal cost of making one more car will be very low.

Solar cells are currently at the cottage industry scale. A *single* nuclear power plant produces 25 times as much electricity in a year as will *all* the solar cells sold in the entire world last year. Of *course* solar cells remain expensive!

Until the solar-cell industry can leap to the next stage—mass production—all the investments the public has made in research and engineering will be left dangling.

What Solar Energy Can Learn from Computer Chips

The solar industry should learn from the experience of the electronics industry. Consider, for example, the history of the integrated circuit. In 1961, Texas Instruments began producing integrated circuits for small, specialized applications. The earliest versions were very expensive. They cost $100 but replaced just a couple dollars' worth of larger electronics, perhaps two transistors and three resistors. There was essentially no market for such devices in the private sector. Other electronics companies sneered at them.

But the American military recognized the potential importance of small, lightweight, low-power integrated circuits. The Department of Defense began to purchase integrated circuits in large quantities. Following a learning curve, the price fell dramatically. As the price fell, numerous private market niches opened up. Soon, the cost of integrated circuits fell to the point where private markets developed explosively.

Sales of Integrated Circuits

Year	Price	Military Percentage
1962	$50.00	100 %
1963	$31.60	94 %
1964	$18.50	85 %
1965	$ 8.33	72 %
1966	$ 5.05	53 %
1967	$ 3.32	43 %
1968	$ 2.33	37 %

In just six years, the price of integrated circuits plummeted 95 percent and an enormous commercial market developed.

In 1971, Intel introduced the first central processing unit (CPU). The early CPUs were not immediately transformational. They were too simple to power anything beyond a calculator. However, those calculators rapidly became more and more sophisticated. By the mid-1970s, microprocessors were performing computations for small computers.

Continuing price drops could then be driven by private demand. The processing power of CPUs has continued to double every eighteen months (as famously predicted by the chairman emeritus of Intel, Gordon Moore). Accompanied by falling costs, this led directly to the information revolution.

Integrated circuits are now dirt cheap and CPUs are ubiquitous in our homes, cars, and workplaces. But if the government had not purchased huge quantities of chips before they were cheap enough for commercial applications, the technology might never have become cost-effective. The information revolution would have been delayed, perhaps indefinitely.

Buy Solar

The same basic economic approach would work for solar cells (which, coincidentally, are made of the same semiconductor materials as computer chips). Every time the volume of solar cells has increased in the past, the cost has fallen. Convincing evidence shows that this learning curve will continue as it has with computer chips. The consumer market already is there, just waiting for the price of solar cells to drop. Every poll ever conducted concludes that most Americans would prefer to get their energy from the sun, if only they could afford to.

Put simply, a national program to spend $5 billion over the next four years could make solar cells commercially viable for a significant portion of all new electrical generation worldwide. An announcement that the government will buy $1 billion worth of solar cells in 2001 at (an inflation-indexed) price of $3.50/watt; $1 billion worth in 2002 at $2.50/watt; $1 billion worth in 2003 at $1.50/watt; and $2 billion worth in 2004 at $1.00/watt would catapult America back into the global lead in solar energy. (If the new president makes it a priority, the entire program could be completed in his first term.)

The economic consequences of a program designed to reduce the price of solar cells would all be positive. Such an initiative would open a large new global market for American technology and create a rich new source of American jobs.

When the price of a product is lowered due to bulk orders from the government, the change is permanent. Unlike, say, taxes and tax credits, economies of mass production cannot be repealed. Part of the beauty of this approach is that, after just a few years, as with computer chips, no additional government subsidy is required.

The program is risk-free. If no company steps forward to sell the equipment at those prices, the program will not cost a cent. But if some companies are able and willing to meet the challenge, the impact on the world will be revolutionary. 🌍

Make That a Grande: CAFE Standards

CAFE standards, like democracy, are the worst system— except for all other possible systems.

In 1975, Congress passed a law to increase automobile fuel efficiency by establishing corporate average fuel efficiency standards—referred to as CAFE (pronounced ka-FE) standards. These require each manufacturer to achieve a certain efficiency level for all the car models in its total fleet, averaged together as a whole.

As discussed earlier in the section on Clean Transportation, adroit lobbying by Detroit has kept CAFE standards from being strengthened for the past 15 years. The CAFE standard for cars remains stalled at 27.5 miles per gallon, and 20.7 miles per gallon for pickups, vans, and sport utility vehicles. In 1999, the mileage of the combined truck and car fleet fell to its lowest in 20 years—23.8 miles per gallon. Nevertheless, in October 1999, Congress once again placed a provision in the transportation appropriation that prevents the federal government from even studying the possibility of raising fuel-efficiency standards for new cars and light trucks. After complaining about the provision, President Clinton signed the bill into law.

In addition to the weaker standards for SUVs, CAFE standards are riddled with many other loopholes. For example, if an automaker exceeds its CAFE requirements one year, it can apply its surplus to the next three years to help balance worse averages. Even more disturbing is that it can take credit today for efficiencies it hopes to achieve in the future!

Automakers manipulate model-year designations shamelessly. For example, General Motors declared that the 1998 model year would *end* in January 1998 for its Chevrolet Tahoe, GMC Yukon, and Chevy/GMC Suburban sport utility vehicles. The 1998 model year had begun just four months earlier. Although the SUVs were not changed in any way, they were designated 1999 models in February of 1998.

Having bent the rules past all recognition, automobile manufacturers now plan to pulverize them by introducing as many as two dozen light trucks that are not trucks at all, but cars. The first of these will be DaimlerChrysler's retro-styled PT Cruiser, based loosely on the Dodge Neon. The Cruiser is no more a truck than is a Honda Civic, but Chrysler contends that the 30 mile-per-gallon vehicle should serve to raise the average mileage of its trucks, vans, and SUVs.

Environmentalists favor dramatic increases in the fuel efficiency of the vehicle fleet. The single most effective thing to do would be to raise the price of gasoline to reflect its true costs: the cost of keeping an American naval fleet in or near the Persian Gulf, the cost of coping with global warming, the health costs of urban air pollution, and so forth. Europe and Japan have raised the price of gasoline to reflect such hidden costs. As a result, cars are much more efficient in Europe and Japan, and cities are designed to decrease reliance on automobiles.

Americans, however, practically consider it a constitutional right to have gasoline that is cheaper than bottled water. Proposing a mere four-cent gasoline tax is viewed as a profile in political courage.

With gasoline now cheaper than it has been for 75 years (after adjusting for inflation), Detroit has difficulty selling enough fuel-efficient cars to raise its fleet averages to meet the CAFE requirements. One reason Detroit puts so much effort into shifting its big-car customers out of luxury cars and into luxury SUVs is that—as noted above—SUVs fall under a wholly separate CAFE standard. (This separate-CAFE-standard loophole is so big you can drive a giant Ford Excursion through it.)

Gasoline taxes are always defeated by an unholy alliance between Sun Belt politicians who are bought and paid for by the oil industry and liberals who have a knee-jerk reaction because gasoline taxes are regressive. (There is not a strong correlation between earning a lot of money and using a lot of gasoline.) Such liberals are short-sighted. Gasoline taxes are a huge potential revenue source, and part of that revenue could be used to reduce other taxes, like payroll taxes, that are much *more* regressive.

Despite their flaws, CAFE standards are the most effective instrument we can get through Congress for holding Detroit's tires to the fire. In lieu of a European-style gasoline tax, we need strict, enforceable standards that will force a new generation of technology—hybrid cars and fuel cells—swiftly onto the market.

The Earth Day Clean Energy Agenda (see page 162), calls for combining the CAFE standards for cars and light trucks into just one category, and raising that standard to 45 miles per gallon by

2010, and 65 miles per gallon by 2020. You can eliminate Detroit's most effective argument against this goal by buying the most fuel-efficient vehicle in the size category you need (and, if it meets your needs, buying a hybrid). 🌍

Strengthen Appliance Efficiency Standards

Appliance efficiency standards have been a stunning success.

Responsible manufacturers like Whirlpool tried for years to sell more efficient refrigerators, but could not get consumers to look beyond the initial price. In part, this was because consumers are skeptical of manufacturer's claims. They could not believe that a product that looked identical to a competing product would actually save much electricity. Whirlpool lost sales to competitors whose much-less-efficient models were slightly less expensive.

Energy-efficient refrigerators cost a little more to make. They have better door seals, more insulation, and efficient compressors. They typically recapture this extra expense in four to six months through lower electric bills. But most new refrigerators are purchased by someone other than the person who will be paying the electric bills. For example, contractors who build spec housing or rental apartment buildings buy refrigerators in carload lots. The efficiency of the refrigerator is not one of the hundred top concerns of a consumer considering a new-construction home or a young couple renting an apartment. The builder therefore has no incentive to pay an extra $50 each for thousands of refrigerators.

When regular folks buy a new refrigerator, it is almost always because their existing refrigerator suddenly died. Their food is spoiling. They are facing a major, unanticipated expense. If they can see a way to save $50 on the initial purchase price, they grab it.

Under the leadership of the Natural Resources Defense

Council, some leading appliance manufacturers teamed with appliance retailers, gas and electric utilities, environmental groups, and state energy offices to pursue energy efficiency. The result was the National Appliance Energy Conservation Act of 1987. Manufacturers accepted that the Department of Energy would regularly review appliance efficiency standards and propose new ones to achieve "the maximum level of energy savings that is technologically feasible and economically justified." The standards are also forbidden by law to compromise consumer convenience or eliminate features.

Through savings on home electricity bills, current appliance standards will result in net savings to consumers of $200 billion over the lifetimes of the next generation of appliances. That equals $2,000 in after-tax dollars for every U.S. family, a sum that dwarfs most of the tax cuts that Congress has considered in the last quarter-century. In addition to saving boatloads of money, appliance efficiency standards conserve enormous amounts of energy and reduce warming gases and other pollutants.

Yet powerful members of Congress invariably fight every new efficiency standard that is proposed, on the grounds that such standards represent government tinkering with the sanctity of the marketplace. New standards are routinely delayed for years as congressional committees bury them in paper. Tell your representative and senators loudly and clearly that you support programs to squeeze the wasted energy out of appliances. 🌎

Kyoto, or Bust

"To cherish what remains of the Earth and to foster its renewal is our only legitimate hope of survival."
—*Wendell Berry*

Officials from more than 150 countries met in Kyoto, Japan, in December 1997, to negotiate a treaty that would steer the world away from global warming. The resulting Kyoto Protocol on Global Warming requires industrialized countries to reduce their emissions of warming gases. The United States, for example, would have to reduce its global warming pollution 7 percent below 1990 levels by 2008-2012. Because U.S. emissions of such gases have increased about 12 percent since 1990, we would need to reduce our current emissions about 19 percent—by roughly one-fifth—from today's levels.

Although not an effortless undertaking, reducing current emissions is much more attainable than you might expect. The obstacles are mostly political. Existing technology can meet the Kyoto targets in a cost-effective manner, if supported by a legal and political framework that is not weighted to give extraordinary advantages to carbon-based fuels. However, the oil and coal industries, together with their allies in oil-exporting countries such as Saudi Arabia, have shown they will not surrender their favored positions without a fight to the death.

At the time of this writing, the heads of state of 84 countries have signed the Kyoto Protocol, but the legislatures of only 14 countries—all from the developing world—have ratified it. To become legally binding, 55 countries must ratify it.

According to the October 19, 1999, *Wall Street Journal*, "Without support from the U.S., the world's largest producer of greenhouse gases, the Kyoto pact in its current form would be dead in the water . . . . Uncertainty in the U.S. has set off a global game of chicken. Companies in Europe and Japan won't carry out emissions-cutting plans if they see their U.S. counterparts withholding political support and investments in cleaner technologies . . . ."

As of late 1999, both Democratic candidates for the presidency have said they support the Kyoto Protocol, will submit it to the Senate for ratification, and will seek to build public support for it. None of the Republican candidates has made a similar commitment.

International agreements must be ratified by two-thirds of the United States Senate. Unfortunately, the quality of dialogue on this issue in the Senate is a testament to the poor quality of science education in America. It is doubtful that the treaty would receive the support of even one-third of the Senate, much less two-thirds.

Without a public uprising, the United States Senate almost

certainly will not ratify the treaty. Every candidate running for election to the United States Senate should be grilled at every campaign stop from now to the election on whether he or she will vote to ratify the Kyoto Protocol on Global Warming. Arrange for different questioners to nail the candidate at every public meeting. Try to place op-ed pieces, or at least letters to the editor, in every newspaper in the state, including small weekly papers.

If a U.S. Senate candidate in your state will not forthrightly pledge to vote to ratify Kyoto, learn who is contributing hard and soft money to his campaign. Learn which groups are supporting his candidacy with so-called "independent issue campaigns"—one of the many giant loopholes in the laws regulating campaign financing. Web sites such as *www.publicampaign.org.* will steer you to sources of solid information.

Global warming is an issue affecting the future of life on our planet. Work to make sure it gets at least more attention than fixing potholes. Make sure it's not an issue that can be artfully dodged with slick political rhetoric. The question is: "Will you vote for ratification of the Kyoto Protocol on Global Warming?" The correct answer is: "Yes." No other answer is satisfactory.

If you have a candidate who favors ratification, know that big oil money will be flowing to his opponent. So volunteer, make contributions if you can, walk precincts, put up yard signs, make phone calls, pass out bumper stickers, and do whatever you can to help level the playing field.

If half the incumbents who oppose ratification lose their seats in the 2000 elections—and if there is a clear connection between their stand on Kyoto and their new unemployment—their like-minded colleagues will get the message. The only thing in American politics that speaks more loudly than money is a riled-up citizenry.

So get riled up! 🌍

Nine Ways to Kyoto

Once we no longer live beneath our mother's heart, it is the earth with which we form the same dependent relationship, relying . . . on its cycles and elements, helpless without its protective embrace.—Louise Erdrich

After the United States ratifies Kyoto, we will need to take concrete steps to implement the treaty. Fortunately, we have the scientific excellence and the policy know-how to rather easily meet the Kyoto requirements. Unlike war, racism, inequality, and other sobering crises that confront us, we know how to solve the energy crisis.

The real obstacles to implementation are the oil, gas, and coal companies. Endowed with ample money and too much political power, they have fought successfully to thwart progress and feather their own nests. Some politicians appear to be little more than tools for the carbon fuel cartel. For example, some members of Congress argue that compliance with Kyoto will turn the United States into a third-world country. This is a downright silly claim, even for a politician.

Done carefully, Kyoto compliance will barely be noticed. Moreover, where it is noticed, its impacts on the economy, public health, and the consumer's pocketbook will be almost entirely beneficial.

The nine policies I list below would achieve the Kyoto targets with substantial economic net benefits. I've ranked them according to the rough percentage of Kyoto goals each would represent if put into practice.

Make More Efficient Cars: 22 Percent.
Between 1975 and 1984, tough corporate average fuel efficiency (CAFE) standards caused American cars to double their miles per gallon. However, for the last 15 years, Congress has refused to raise the CAFE standards, and the fuel efficiency of the American vehicle fleet has deteriorated. Increasing the average fuel economy of new cars and light trucks in a single

combined category to 42 miles per gallon by 2010 would reduce U.S. carbon emissions by 109 million metric tons. To help build the market, the gas guzzler tax should be converted into a fee and rebate system, whereby buyers of inefficient cars and trucks subsidize those who purchase super-efficient vehicles.

Create a National Utility Trust Fund: 20.5 Percent.

Historically, the utility industry has funded programs to boost energy efficiency, develop renewable power sources, help low-income families, and promote other public benefits. However, deregulation and the resulting competition have led to a decline in such expenditures for the last five years. A federal trust fund proposed by Senator Jeffords and Representative Pallone would restore much of this public benefit funding. The cost to the average consumer would be about $1 per month. It would reduce carbon emissions by 70 million tons per year in the residential sector, and another 33 million tons in the commercial/industrial sector, for a total of 103 million tons per year by 2010.

Create Incentives to Reduce Industrial Energy Use: 14 Percent.

The industrial sector accounts for about 40 percent of U.S. energy use. Two-thirds of this is in the manufacturing sector, with six industries accounting for most of the energy use. Opportunities for reductions of industrial carbon emissions are stunning. Germany, the Netherlands, and Denmark have pioneered strategies to encourage industries to reduce greenhouse gas emissions. Based on their experience and upon detailed analysis of similar opportunities in the United States, a combination of technical and financial assistance can lead to an annual reduction in industrial carbon emissions of 71 million tons per year by 2010.

Establish a Federal Renewable Portfolio Standard: 11 Percent.

As part of the price of deregulation, states have been requiring utilities to obtain a certain amount of future electricity from solar, wind, geothermal, and biofuels. A federal bill authored by Senator Jeffords would set a standard of obtaining 10 percent of America's electricity from non-hydro renewable sources by

2010. This measure would raise consumers' electric bills less than $0.50 per person per month, while cutting carbon emissions by 55 million metric tons per year.

Encourage Combined Heat and Power: 10 Percent.

In a typical power plant, 70 percent of the energy in the fuel is thrown away as waste heat. When power plants are located at facilities that can use this heat, it is no longer a waste but a valuable new energy source. Such combined heat and power plants produce just 9 percent of U.S. electricity. Dramatic recent improvements in gas turbines, microturbines, and fuel cells make these distributed power sources competitive on a level playing field. However, in most states the game is rigged in favor of the utility and against the small power producer. A handful of reasonable state and federal policy changes to eliminate this utility-sponsored discrimination would lead to explosive growth in this field—doubling the amount of power generated from such plants, and reducing the amount of carbon emitted by 49 million metric tons.

Close the Power Plant Emissions Loophole: 9 Percent.

Almost 1,000 old coal-fired power plants, representing 300,000 MW of generating capacity, are exempt from today's air pollution standards because they were grandfathered out of complying with the Clean Air Act. When the legislation was passed, everyone thought these ancient plants would be closed by now. But deregulated utilities are operating them long beyond their design life because of their low cost. In late 1999, the EPA filed lawsuits against many of these plants to try to force them into compliance. Making all power plants meet the nation's clean air standards would decrease their emissions of harmful pollutants by 75 percent—with great public health benefits. It would also speed up the replacement of these dinosaurs with clean, modern plants, eliminating at least 43 million metric tons of carbon per year by 2010.

Build Better Buildings: 4.5 Percent.

Building codes are arcane and boring, but they have enormous impacts on the amount energy used to heat, cool, light, and power buildings. The federal Energy Policy Act of 1992 requires all states to adopt commercial and residential building codes. Many haven't. Model building codes should be updated to reflect current cost-effective efficiency standards, and the Department of Energy should be given authority to see that such codes are adopted and followed by all states. Coupled with new programs to promote efficiency retrofits of existing buildings, these measures would reduce carbon emissions by 25 million tons per year in 2010.

Manufacture Efficient Appliances: 4.5 Percent.

Existing appliance efficiency standards will reduce the nation's yearly emissions of carbon by 65 million metric tons by 2010. They will save as much energy as 90 huge coal-fired power plants could produce—and at the same time provide consumers a net savings of $160 billion. Congress has made it virtually impossible for the Department of Energy to issue the efficiency standards it is required by law to develop. Consequently, the program is many years behind schedule. Tough new efficiency standards should be set for fluorescent lighting ballasts, water heaters, clothes washers, central air conditioners, heat pumps, transformers, refrigerators, furnaces and boilers, commercial air conditioners, gas ranges, and reflective lamps. Energy Star labeling programs should also be expanded to a wider range of products. This decision to implement current laws would save an additional 23 million tons a year by 2010, about 4.5 percent of our Kyoto target.

Set Greenhouse Gas Standards for Vehicle Fuels: 4.5 Percent.

Just as a renewable portfolio standard for electric utilities reduces the amount of carbon emissions from electricity each year, greenhouse gas standards could be used to gradually reduce the net amount of carbon emitted each year from motor vehicles. This would create immediate incentives for cellulosic ethanol and, soon, for renewable hydrogen. A 5 percent reduction in carbon emissions from motor vehicle fuels by 2010 would eliminate 22 million metric tons per year.

Overlapping All Categories: Establish Tax Incentives for Smart Energy.

Far from turning America into a second-rate nation, this set of policies is designed to make a substantial positive contribution to the economy. It would require a total investment of $213 billion but it would save $416 billion in fuel and operations and maintenance—for a net savings of $203 billion. If the program were continued through 2020, the net savings would grow to $510 billion. The net savings could have been far greater if we had started sooner, and (with 2010 barreling toward us) success will rapidly get more expensive. Eventually it will become impossible at any price.

The government can help commercialize products that promote the Kyoto goals by procuring those products in large quantities (thus reducing their prices) and by giving them favorable tax treatment. Energy taxes can be easily made revenue-neutral. With revenue-neutral taxes, people pay no more taxes overall, but their taxes are lowered on things we want to encourage (like labor and saving) and raised on things society wishes to discourage (such as burning carbon fuels). The cost to society of a revenue-neutral tax is zero.

Policies have consequences. If the United States still used as much energy per unit of Gross Domestic Product as in 1970, we would have consumed 153 quadrillion BTUs this year instead of 95. This 68 quadrillion BTU saving from investments in efficiency is a testament to our ability to choose our fate. If even half of this saving had not occurred, the nation's energy bill over the last 30 years would have been $2.5 trillion higher, and we would have emitted 8 billion more tons of carbon. No single policy brought about this improvement. Rather, scores of policies influenced the behavior of millions of individuals and enterprises.

There is no single magic bullet, either in the form of a perfect new technology or a comprehensive new policy, that will solve global warming. But the new technologies and policy innovations discussed in this book can each take a bite out of the problem and make it more digestible. 🌎

142

Extended Warranty: Resources and Activism

This is the section for activists. It begins with some background on how Earth Day itself started during the heyday of American activism and grew over the years. You don't have to chain yourself to a tree to be an activist.

Your "acts" may be as simple as signing a petition, boycotting a product, supporting an environmental group, and voting for green candidates. We make it easy by providing petition forms, lists of bad companies, and resources for learning about good politicians.

A Brief History of Earth Day

When Senator Gaylord Nelson, chair of the first Earth Day, encouraged me to drop out of graduate school at Harvard in the fall of 1969 to become Earth Day's national coordinator, I viewed the world this way: America was growing wealthier, but the quality of life was getting worse. There was little connection between the way economists measured progress and the things people really cared about. There was a growing gap between what statisticians counted and what really counts.

Robert Kennedy—the figure who more than any other lured me into political activism—summed it up as follows:

> The gross national product includes air pollution and advertising for cigarettes, and ambulances to clear our highways of carnage. It counts special locks for our doors, and jails for the people who break them. The gross national product includes the destruction of the redwoods and the death of Lake Superior. And if the gross national product includes all this, there is much that it does not comprehend. It does not allow for the health of our families, the quality of their education, or the joy of their play. It is indifferent to the decency of our factories and the safety of our streets alike. It does not include the beauty of our poetry or the strength of our marriages, the intelligence of our public debate or the integrity of our public officials. It allows neither for the justice in our courts, nor for the justness of our dealings with each other. The gross national product measures neither our wit nor our courage, neither our wisdom nor our learning, neither our compassion nor our devotion to country. It measures everything, in short, except that

which makes life worthwhile; and it can tell us everything about America—except whether we are proud to
be Americans.

The message of that first Earth Day, April 22, 1970, was broadly
inclusive. Without compromising any core values, Earth Day organizers consciously set out to pull together all kinds of people.

Earth Day organizers in different cities variously protested
against air pollution, oil spills, vanishing wilderness, eight-lane
freeways cutting through their neighborhoods, DDT and lead
poisoning in America's ghettos, and dozens of other specific
blights. But Earth Day was not *just* a collection of piecemeal
protests. Earth Day was in effect a huge town meeting, asserting
that there is more to the Amer-
ican Dream than ever-increasing consumption.

Until 1970, economists had
completely dominated most
domestic-policy debates with
theories about the impact of
various policies on economic
growth, capital formation,
jobs, productivity, and the like.
Arguably, Earth Day's most
important achievement was to
introduce biology forcefully
into public policy. Earth Day
stood for the proposition that
the American Dream could not
be sustained unless it reflected
ecological principles.

> ### ENERGY EXTRAS
>
> *In today's era of
> jobs-versus-owls
> bumper stickers, few
> remember that
> organized labor was
> the largest source of
> funding for the first
> Earth Day. Enlight-
> ened labor leaders
> understood that facto-
> ries were the most
> polluted places in
> America.*

Not everyone saw it that way. Maurice Stans, then Secretary
of Commerce,denounced Earth Day as the latest wave of wild-
eyed radicalism by kids who had lost touch with reality. The
Daughters of the American Revolution found it sinister that
Earth Day happened to fall on Lenin's birthday.

But for most Americans, the Earth Day message smacked of
simple common sense. Earth Day staff members put a huge

ENERGY EXTRAS

Kurt Vonnegut spoke for most of us at the first Earth Day when he said of President Nixon, "I am sorry he's a lawyer; I wish to God that he was a biologist. He said the other night that America has never lost a war, and he wasn't going to be the first American president to lose one. He may be the first American president to lose an entire planet."

amount of effort into enlisting schools, colleges, churches, garden clubs, civic organizations, businesses, and laborers. A broad cross-section of America came together around a common set of values, prompted perhaps by the most basic biological drive—the drive for survival.

Aided by stunning NASA photographs of the Earth from space, Earth Day's organizers argued that we needed to pay more attention to the web of life and our role in it. Human well-being, we said, is linked more closely than most people realize to the great marine and terrestrial ecosystems. Carbon emissions must not exceed nature's capacity to withdraw carbon dioxide from the atmosphere and fix the carbon. Logging cannot exceed the rate of tree growth, and must be sensitive to the needs of complex forest ecosystems. Fish catches cannot exceed sustainable yields, and spawning habitats must be protected. Soil erosion must not be faster than soil formation. Water must not be pumped out of aquifers faster than they recharge. *This is all common-sense stuff!*

Earth Day launched the modern environmental movement. From 1970 to 1974, the United States passed a wave of environmental legislation, comparable in some ways to the New Deal in its sweep and impact on the way America does business: the Clean Air Act, the Clean Water Act, the Endangered Species Act, the Safe Drinking Water Act, the Federal Coastal Zone Management Act, and so forth. Environmentalists also put muscle behind the creation of the Environmental Protection Agency and helped pass the Occupational Health and Safety Act as an effort to eliminate in-plant pollution.

Environmentalists are sometimes incorrectly caricatured as doomsayers. Much environmental literature *does* focus upon disasters that may come if we don't mend our ways. *Silent Spring*, the first influential modern environmental book, described an unbearably bleak future—in the hope of stirring up changes that were needed to keep it from coming to pass. In truth, environmentalism is probably the most hopeful movement to develop during the twentieth century. If there were no hope, we wouldn't bother. As the history of Earth Day shows, environmentalists can successfully overcome formidable odds with hard work, creativity, good science, and smart politics.

Now it's your turn to grab the baton. 🌍

Boycott Global Climate Coalition Companies

The Global Climate Coalition (GCC) is the biggest political obstacle in the United States to efforts to halt global warming. The GCC was founded in 1989 to counter the "myth" of global warming. Its members represent coal, oil, chemical, and automotive interests.

The GCC's denial of reality follows a path blazed by previous industrial campaigns of misinformation. The tobacco industry, for example, continued to deny that cigarettes caused cancer even after every educated person in the nation knew this assertion was a bald-faced lie. Similarly, the GCC is waging a vigorous, expensive propaganda campaign on behalf of the fossil-fuel industry, brazenly proclaiming that there is no problem despite a scientific consensus to the contrary.

When twenty-first century historians reflect back upon the causes and the full impact of global warming, the GCC will be judged harshly. Placing short-term profits above concern for humanity, the GCC's policies are likely to cause far more death and misery than those pushed by the tobacco industry.

The GCC's propaganda should not be confused with the genuine scientific debate about the pace and magnitude of future changes in global climate. This scientific debate is healthy. There is room for honest disagreement about the implications of climate change for Caribbean hurricanes, Chinese floods, African droughts, Siberian permafrost, and Nebraskan corn production. Although there is a strong consensus that the planetary changes caused by fossil-fuel consumption are unprecedented and potentially catastrophic, we don't know (and we can't know) in great detail what will happen to Earth over the next hundred years. The scientific process, by its very nature, always remains open to competing ideas and explanations.

The Global Climate Coalition shamelessly exploits this scientific debate. It argues that no actions should be taken until the scientific debate is completed. But a consensus over the big issue already exists: global warming is real and it demands action now. The "details" of climate change will not be settled until large-scale global warming, irreversible on a human timescale, has already happened.

In other words, the GCC demands that we not act until it is too late.

As public policy, this is absurd. But the GCC is rich, powerful, clever, and effective. When the full memberships of the trade associations belonging to the GCC are added together, this powerhouse can claim to represent more than 230,000 companies! When they speak, a lot of politicians listen.

Perhaps the GCC's most reprehensible coup was its nimble playing off of rich against poor at the Kyoto Climate Change conference. The GCC pointed out to the U.S. Congress that emissions of greenhouse gases were growing more rapidly in developing countries than in the industrialized world. (This is natural, because they are beginning from such a very low per-capita base.) The GCC told Congress that if the U.S. limited its CO_2 before the developing world made similar commitments, it would be futile. All nations, rich and poor, needed to proceed together.

At the same time the GCC was meeting with leaders of

developing countries and urging them not to implement the Kyoto accords until they were more fully developed. The industrial world was built on cheap fossil fuels, it said, and now the North wants to keep the South in permanent second-class status. If America and Europe, with their gluttonous energy consumption, are genuinely worried about fossil fuels, let them tighten their own ample belts before asking you to make sacrifices.

Even as his lobbyists were vigorously urging Congress not to support any international global warming agreement that did not require immediate CO_2 reductions by China, the CEO of Exxon was telling the World Petroleum Congress *in Beijing* that increased fossil-fuel use in China is essential both for economic growth and for the elimination of poverty, which is the worst polluter. He continued, "The most pressing environmental problems of the developing nations are related to poverty, not global climate change. Addressing these problems will require economic growth, and that will necessitate increasing, not curtailing, the use of fossil fuels."

As a result of the GCC's sleight-of-hand, the U.S. Senate voted 95 to zero not to ratify Kyoto until key parts of the developing world were included. Similarly duped, much of the developing world refuses to proceed until it sees evidence of good faith by the United States.

The fossil-fuel industry, in short, is behaving just like the cigarette industry. It is doing everything it can think of to keep the world hooked on the product it sells—in this case, an ever-increasing dose of carbon-based fuels. Like most addictions, this feels good for a while, but it is ultimately ruinous.

The GCC's campaign should earn its members the only reproach they care about, a refusal by decent people to do business with them. Everyone who cares about the future of the world should refuse to buy the stock, and office companies should boycott their goods and services. College students should refuse to interview for jobs with them.

Most of the 230,000 firms that the GCC claims as supporters are only distantly or indirectly related to this deviously organized group. The GCC membership is full of trade associations like the American Forest and Paper Association, the American Iron and Steel Institute, the National Association of Manufacturers, the Edison Electric Institute, and the U.S. Chamber of Commerce, which represent hundreds of companies. Many of those companies may not even know that their trade associations are involved in this nefarious campaign. Call it to their attention! If your utility's trade association won't drop out of the GCC, lobby the utility to withdraw from its trade association.

Shell, BP, Arco, and Sun have resigned from the GCC. I try hard to buy gasoline only from these companies, rather than from ExxonMobil, Texaco, and Chevron, which remain staunch GCC supporters. Shell, BP, Arco, and Sun are not corporate candidates for sainthood. However, they have broken ranks with their cohorts and taken a courageous stand on the most important issue facing the carbon-fuel industry.

The Association of International Automobile Manufacturers withdrew from GCC some years ago. Ford Motor Company resigned in December, 1999, and DaimlerChrysler followed a couple of weeks later. All the focus is on pressuring General Motors to follow suit in 2000.

Review the following list of members. Think hard about whether there are ways, as a shareholder or consumer, you can let the member companies know that you are disturbed by their association with the Global Climate Coalition.

A final note: If this effort begins to enjoy success, you will begin to receive communications from GCC companies stating that they understand your position. However, they may explain, rather than leaving the GCC, they have decided to "remain and try to reform it." (Such statements are issued every single time a corporate responsibility campaign begins to have an impact.)

Environmentalists have no interest in seeing the GCC "reformed" by ExxonMobil and the rest, because *the GCC has no legitimate mission*. Its only reason for existence is to promote energy policies that harm the earth. And any company that remains a member is supporting something very wrong.

Global Climate Coalition Members

Board membership:

Allegheny Power
Ameren Services
American Farm Bureau Federation
American Forest & Paper Association
American Iron & Steel Institute
American Petroleum Institute
American Portland Cement Alliance
Association of American Railroads
Chemical Manufacturers Association
Chevron
CSX Transportation, Inc.
Dresser Industries
Drummond Company
Duke Energy
Edison Electric Institute
ExxonMobil
General Motors
Illinois Power Company
IPSCO Steel Inc.
National Association of Manufacturers
National Lime Association
National Mining Association
National Rural Electric Cooperative Association
Norfolk Southern
Parker Drilling Company
Rail Progress Institute
Southern Company
TECO Energy, Inc.
Texaco Inc.
U.S. Chamber of Commerce
USX Corporation
Virginia Power

General membership:
 American Plastics Council
 BHP Minerals
 Bethlehem Steel
 Consumers Energy
 Eastman Chemical
 First Energy
 Hoechst Celanese Chemical Group
 Kaiser Aluminum & Chemical Corp.
 Northern Indiana Public Service Co.
 Pennsylvania Power & Light
 Santa Fe International Corporation
 Union Pacific 🌍

Elect Green Candidates

The environmental movement has been bipartisan ever since it began in this country. Republican Theodore Roosevelt tripled the size of the national forests, expanded the national parks, and established the first national monuments. Republican Richard Nixon took the initiative to establish an Environmental Protection Agency, and he named strong leaders to head it.

We remember Ed Muskie as the father of the Clean Air Act and the Clean Water Act. But Republican Senator John Sherman Cooper (the ranking minority member of the Muskie Subcommittee on Air and Water Pollution of the Senate Public Works Committee) co-directed that bipartisan effort. Senator Cooper was creative and committed, and he brought most of his Republican colleagues along with him.

If this bipartisan support had endured, the environmental movement might have been indifferent to the fates of the two parties. However, to the *double* misfortune of environmentalists, the condition of our planet has become an increasingly partisan issue. On one hand, the new Republican leadership is senselessly hostile, and some of its key chairmen are blithely doing irreparable damage to the Earth. On the other hand, the Democrats now take environmentalists for granted. Because the

---

E N E R G Y   E X T R A S

John Ehrlichman, head of the Domestic Council under President Nixon, once confided to me how Nixon had decided to create the EPA. This decision was not motivated by a deep presidential sensitivity for the environment. Indeed, Nixon thought environmentalism reflected a "wimpy reluctance to pay the price of progress." Still, Nixon considered Senator Ed Muskie to be his likely Democratic opponent in the 1972 election. He was convinced that Muskie would tout his strong leadership in the fight against air and water pollution. To Nixon, the 20 million people who took part in the first Earth Day bespoke potential political power. He felt he needed to do something to burnish his own environmental credentials. A routine task force recommendation for an Environmental Protection Agency just happened to be at hand.

Republican leadership is so astonishingly anti-environmental, Democratic leaders feel they can safely ignore us because we are a captive constituency with no place else to go.

There are at least three good reasons why environmentalists should *not* write off the GOP:

(1) Democrats have no monopoly on good ideas.

(2) All politicians ignore any constituency they can take for granted.

(3) Environmentalists have a far broader base of support than the Democrats.

Is this realistic? Is there any chance the Republicans can ever again be environmental allies? Well, was there a chance that the racist, segregationist Democratic Party of the first half of the twentieth century could become the party of civil rights? That it could win African-Americans away from Abraham Lincoln's party?

ENERGY EXTRAS

It is impossible to exaggerate the importance of reaching out broadly for environmental support. The genius of the original Earth Day lay in the way it enlisted all segments of American society. Earth Day's organizers had no interest in events where the Sierra Club just talked to the National Audubon Society. Instead, we invited students, teachers, clergy, businesses, laborers, farmers, ethnic minorities, and civic organizations to help plan the campaign. Our largest single donor turned out to be the United Auto Workers—for a campaign that attacked automobile pollution.

Party politics is a matter of leadership and definition. I would not be shocked to see an innovative brand of environmentalism emerge again from Republicans such as Bill Ruckelshaus, Russell Peterson, Bill Weld, and Jim Jeffords. Some of their policies are sure to be different than those of Barbara Boxer, John Kerry, and Bruce Babbitt. But, like the green pricing proposals of the late Republican Senator John Heinz, they may offer creative new opportunities to cut through the current political logjam.

In a nation emerging from the polarizing 1960s, Earth Day provided a vehicle in which diverse people could come together to promote their visions of a better future. It introduced a value system that still commands the support of 80 percent of all Americans. It launched the modern environmental movement as the owner of the ultimate big tent.

Many well-educated environmentalists have a naïve view that they live under a system of government where good science automatically leads to good policy. We don't. We live in a democracy. Votes must be courted and won.

Not one politician in a thousand cares as much about what the Society for Conservation Biology or the National Academy of Sciences say about an issue as what the Chamber of

Commerce, the AFL-CIO, the AARP, and the Christian Coalition say. These groups have a history of delivering money and votes. What the environmental movement needs most is a green political army, able and willing to do the hard work of winning in a democracy.

What Is at Stake?

The stakes in the current political battle are far greater than we may be able to conceive. American elections in the early years of this millennium will be comparable to the battles that surrounded the birth of Gutenberg's information revolution and the subsequent Reformation, Enlightenment, scientific revolution, and birth of democracy. Those earlier struggles could very easily have been lost, leading to a continuation of the Western theocratic era, with its Inquisition to root out any who questioned the accepted wisdom.

The same political forces that are uniting to squash environmental protection are seeking to defund science, defund the arts, and expand censorship. They seek to consolidate control of Congress and the White House in archconservative hands for the foreseeable future. The current congressional leadership has pulled the pendulum so far to the right that they make Calvin Coolidge look like a bleeding heart. Political pendulums swing back only when enough people pull on them. It is time to re-strike the balance.

Go to the League of Conservation Voters Web site and review the voting record of your federal officials. If you don't have a computer, either use one at your local public library or contact the group directly and order a voter's guide:

League of Conservation Voters
1920 L St., NW, Suite 800
Washington, D.C. 20009
202-785-8683
www.lcv.com

Support the League financially. Contribute money to its campaign to defeat legislators who are bought and paid for by the Global Climate Coalition. And let every candidate in your area

---

ENERGY EXTRAS

Eleven percent of U.S. students don't graduate from high school, and three-quarters don't graduate from college. Environmental groups do a good job recruiting college graduates, and even better among those who obtain advanced degrees. That provides a solid, influential base of support. But in a democracy, majorities matter. To be successful in the long run, the environmental movement must put much more effort into winning the hearts and minds of those who don't attend college.

---

know that, in future elections, your votes will go only to politicians—from either party—who share your desire for a safe, healthy, diverse, sustainable environment. 🌍

Earth Day Sponsors

Many national environmental groups are "sponsors" of Earth Day 2000. All these organizations have made a financial contribution to the campaign and lent support in other ways. Some are strategic partners, with energy-related projects that will come to fruition in the spring of 2000. Others are volunteering staff time, office space, technical advice, or other resources to various Earth Day efforts.

Earth Day's sponsors all make a valuable contribution in one or more fields of environmental protection. Please visit their Web sites (each has a hot link from the *www.earthday.net* site), and generously support the groups whose priorities you share.

Alliance To Save Energy
1200 18th St., NW, Suite 900
Washington, D.C. 20036-2506
202-857-0666
www.ase.org

American Forests
910 17th St., NW, Suite 600
Washington, D.C. 20006
202-955-4500
http://www.amfor.org/corp

American Solar Energy Society
2400 Central Ave., Suite B-1
Boulder, CO 80301
303-443-3130
www.ases.org

Center For Environmental Citizenship
1611 Connecticut Ave., NW, Suite 3-B
Washington, D.C. 20009
202-234-5990
www.envirocitizen.org

CERES
11 Arlington St., 6th Floor
Boston, MA 02116-3411
617-247-0700
www.ceres.org

Conservation International
2501 M St., NW, Suite 200
Washington, D.C. 20037
800-429-5660
www.conservation.org

Defenders Of Wildlife
1101 14th St., NW, Suite 1400
Washington, D.C. 20005-5605
202-682-9400
www.defenders.org

EarthJustice Legal Defense Fund, Inc.
180 Montgomery St.
San Francisco, CA 94104-4209
415-627-6700
www.earthjustice.org

EarthVoice
2100 L St., NW
Washington, D.C. 20037
202-778-6149
www.earthvoice.org

Environmental Defense Fund
257 Park Ave South, 16th Floor
New York, NY 10010-7386
212-505-2100
www.edf.org

Environmental Working Group
1718 Connecticut, NW, Suite 600
Washington, D.C. 20009-1163
202-667-6982
www.ewg.org

First Nations Development Institute
The Stores Building
11917 Main St.
Fredericksburg, VA 22408
540-371-5615
www.firstnations.org

Friends Of The Earth
1025 Vermont Ave, NW, Suite 300
Washington, D.C. 20005-6303
202-783-7400
www.foe.org

Humane Society International
2100 L St., NW
Washington, D.C. 20037-1595
202-452-1100
www.hsus.org/international

Humane Society of the United States
2100 L St., NW
Washington, D.C. 20037-1595
202-452-1100
www.hsus.org

League Of Conservation Voters Education Fund
1707 L St., NW, Suite 750
Washington, D.C. 20036
202-785-0730
www.lcvedfund.org

National Audubon Society
700 Broadway
New York, NY 10003
212-979 3000
www.audubon.org

National Environmental Trust
1200 18th St., NW, Suite 500
Washington, D.C. 20036-2513
202-887-8800
www.envirotrust.org

National Parks Conservation Association
1776 Massachusetts Ave., NW
Washington, D.C. 20036
1-800-NAT-PARKS
www.npca.org

National Wildlife Federation
8925 Leesburg Pike
Vienna, VA 22184-0001
703-790-4000
www.nwf.org

Natural Resources Defense Council
40 W 20th St., 11th Floor
New York, NY 10011-4201
212-727-2700
www.nrdc.org

The Nature Conservancy
4245 North Fairfax Drive, Suite 100
Arlington, VA 22203-1606
703-841-5300
www.tnc.org

Ozone Action
1700 Connecticut Ave., NW, 3rd Floor
Washington, D.C. 20009-1134
202-265-6738
www.ozone.org

Population Action International
1120 19th St., NW, Suite 550
Washington, D.C. 20036-3678
202-659-1833
www.populationaction.org

Trust For Public Land
116 New Montgomery, 4th Floor
San Francisco, CA 94105
415-495-4014
www.tpl.org

Union Of Concerned Scientists
2 Brattle Square
Cambridge, MA 02238-9105
617-547-5552
www.ucsusa.org

U.S. Public Interest Research Group
218 D St., SE
Washington, D.C. 20003-1900
202-546-9707
www.uspirg.org

Wilderness Society
900 17th St., NW
Washington, D.C. 20006-2290
1-800-THE WILD
www.wilderness.org

Women's Environment and Development Organization
355 Lexington Ave., 3rd Floor
New York, NY 10017-6603
212-973-0325
Fax: 212-973-0335
www.wedo.org

World Resources Institute
10 G St., NE, Suite 800
Washington, D.C. 20002
202-729-7600
www.wri.org

World Wildlife Fund
1250 24th St., NW
P.O. Box 97180
Washington, D.C. 20077-7180
1-800-CALL-WWF
www.worldwildlife.org

The Earth Day Clean Energy Agenda

A rapid transition to energy efficiency and renewable energy sources will combat global warming, protect human health, create new jobs, and ensure a secure, affordable future. The Earth Day Clean Energy Agenda outlines common-sense steps to mobilize American ingenuity and resources for a rapid, comprehensive transition to renewable energy sources.

Clean Power: In the next decade, increase fourfold the amount of energy obtained from non-hydro renewable sources such as the sun and wind. By 2020, produce at least one-third of the nation's energy from renewable sources, and double the efficiency of energy use in homes, buildings, transportation, and industry.

Clean Air: Clean up our power plants by setting progressively tighter limits on all power plant pollution, including carbon dioxide. Close the loophole that allows old coal-fired power plants to pollute much more than newer plants.

Clean Cars: Hold sport utility vehicles, pickup trucks and minivans to the same air pollution standards as cars. Improve the fuel efficiency of new cars and light trucks to a combined average of 45 miles per gallon by 2010 and at least 65 miles per gallon by 2020. Offer incentives that build strong markets for renewable fuels and for clean vehicles powered by hybrid motors and fuel cells.

Clean Investments: Quadruple federal investments in renewable energy and energy efficiency within five years, and continue this momentum over the long term. Stop spending taxpayer dollars to subsidize the coal, oil, and nuclear industries. Provide adequate resources and job training for affected workers and communities to ensure a just transition to a sustainable energy economy.

I/We endorse the
Earth Day Clean Energy Agenda.

Name:_____

Address:_____

Telephone:_____

E-mail:_____

Please return signed form to:
Earth Day Network
91 Marion Street
Seattle, WA 98104-1441

or

sign-up on EDN's Web site: *www.earthday.net*

**Please photocopy this form and circulate
it to all your friends.**

Join Earth Day Network

The Earth Day Network is a worldwide alliance that organizes annual environmental campaigns on and around April 22 of each year. EDN chooses an annual theme for Earth Day, employs organizers around the world, and coordinates global press coverage.

EDN links more than 4,000 affiliated organizations (the "network") together with a world class Web site, *www.earthday.net*. It provides a brief, free, daily electronic summary of environmental news from around the world (*The Daily Gist*). It also provides educational materials to Earth Day coordinators at more than 80,000 K-12 schools, and it coordinates groups at about 1,000 college campuses.

Earth Day Network depends upon tax deductible contributions from those who support its broad educational mission. Your donation will help us mount a successful campaign around the world.

- -

Earth Day Network
91 Marion Street
Seattle, WA 98104-1441

Here is my contribution in support of Earth Day:

◯ $15   ◯ $25   ◯ $50   ◯ $100   ◯ $250

Name:_____

Mailing Address:_____

City, State, Zip:_____

E-mail Address:_____

_____ I would like to subscribe to *The Daily Gist*.

_____ I would like to be placed on an electronic activist list, to be contacted no more than once a month when action is needed on important environmental issues.

(Contributions can also be made on our secure Web site: *www.earthday.net*.)

Support the Earth Day Foundation

All royalties from the sale of this book will go to the Earth Day Foundation to start building a permanent endowment. If you like the idea of your gift remaining intact, supporting the environmental education of new generations of environmentalists, the Earth Day Foundation will be grateful for your support. A gift to the Earth Day Foundation will help ensure that Earth Day endures, regardless of what political winds may blow or what fashions and fads prevail.

Earth Day Foundation
1011 Boren Avenue, PMB 230
Seattle, WA 98104

Here is my gift to the Earth Day Foundation:

◯ $25 ◯ $50 ◯ $100 ◯ $250 ◯ $500 ◯ $1,000 ◯ $5,000

Name:_____

Mailing Address:_____

City, State, Zip:_____

E-mail Address:_____

_____ I would like to subscribe to *The Daily Gist.*

_____ I would like to be placed on an electronic activist list, to be contacted no more than once a month when action is needed on important environmental issues.

(Contributions can also be made on our secure Web site: *www.earthday.net.*)

Bibliography

American Council for an Energy-Efficient Economy. *Green Book: The Environmental Guide to Cars and Trucks.* Washington, D.C.: ACEEE, 2000. A wide-ranging assessment of the environmental impacts of all the major cars sold in the United States, as well as advice on driving and maintenance to minimize your environmental footprint. See also *www.aceee.org/greenercars.*

American Council for an Energy-Efficient Economy and the Tellus Institute: Howard Geller, Stephen Bernow, and William Dougherty. *Meeting America's Kyoto Protocol Target: Policies and Impacts.* Washington, D.C.: 1999. This is one of the best studies of how to reach Kyoto goals and was an important resource for the section "Nine Ways to Kyoto," found on pages 138–142 of this guide.

John Berger. *Charging Ahead: The Business of Renewable Energy and What It Means for America.* New York: Henry Holt and Company, 1997. This is a highly readable survey of who is doing what where in the renewable energy field. Filled with behind-the-scenes stories of corporate intrigue, *Charging Ahead* puts a human face on the renewable energy business.

Brower, Michael and Warren Leon. *The Union of Concerned Scientists Consumer's Guide to Effective Environmental Action.* New York: Three Rivers Press, 1999. This volume helps the consumer distinguish between which goods and services have major environmental consequences and which ones have trivial impacts.

Brown, Lester et al. *State of the World.* New York: W.W. Norton & Co. This definitive annual publication by the staff of the Worldwatch Institute is now published in dozens of countries in a wide range of languages. Each edition is full of provocative essays and reliable information. Of particular relevance to global warming are several chapters written by Christopher Flavin over the years on energy and climate change.

Butti, Ken and John Perlin. *A Golden Thread: 2500 Years of Solar Architecture and Technology.* Palo Alto, CA, Cheshire Books, 1980. A fascinating history of the application of solar energy to human purposes, rich with photographs and illustrations.

Campbell, Colin J. *The Coming Oil Crisis. Petroconsultants, 1997.* This book's provocative thesis, supported by proprietary data, is that oil-producing countries are inflating the amount of oil they claim to have discovered in order to boost the amount they are allowed to pump each year. If Campbell is correct, oil production will begin to decline around the world within the next ten years.

Gelbspan, Ross. *The Heat Is On.* Reading, MA: Perseus Books, 1998. A Pulitzer-Prize-winning author takes a detailed look at how the oil, coal, and utility industries have conducted their massive campaign to deceive the public about global warming. Gelbspan does an excellent job of updating his material on a Web site: *www.heatisonline.org.*

Hawken, Paul and Amory Lovins and L. Hunter Lovins. *Natural Capitalism.* Boston: Little, Brown & Co., 1999. Paints the clearest, most detailed portrait yet of what a pleasant, productive super-efficient future would look like, and describes how to get there from here.

Hertsgaard, Mark. *Earth Odyssey.* New York: Broadway Books, 1998. Mark spent almost a decade traveling around the world to observe many of the most serious environmental crises of our time. He brings thoughtful insights to a wide range of issues, including nuclear power, climate change, and the spread of the automobile around the world.

Johansson, Thomas B. and Henry Kelly, Amulya K. Reddy, and Robert Williams, Eds. *Renewable Energy: Sources for Fuels and Electricity.* Washington, D.C.: Island Press, 1993. For a general audience, this is the best technical introduction to a wide range of renewable energy technologies. The chapters are authored by leading researchers from around the world.

Maycock, Paul D. and Edward N. Stirewalt. *A Guide to the Photovoltaic Revolution: Sunlight to Electricity in One Step.* Emmaus, PA: Rodale Press, 1985. Fifteen years after its publication, this remains the best popular introduction to the science and application of solar cells.

Office of Technology Assessment of the U.S. Congress. *Renewing Our Energy Future.* Washington, D.C.: U.S. Government Printing Office, 1995. OTA-ETI-614. An excellent review of the state-of-the-art in the United States of most renewable energy technologies. Its publication was one of the final acts of the bipartisan Congressional Office of Technology Assessment before Newt Gingrich abolished Congress's science office.

Perlin, John. *From Space to Earth: The Story of Solar Electricity.* Aztec Publications, 1999. Perlin has authored the same sort of excellent, detailed history for solar photovoltaics that he and Ken Butti did in *A Golden Thread* for solar architecture.

Schneider, Stephen, Editor in Chief. *Encyclopedia of Climate and Weather.* New York: Oxford University Press, 1996. The first source book to turn to for everything you might want to know about climate.

Smeloff, Ed and Peter Asmus. *Reinventing Electric Utilities.* Washington, D.C.: Island Press, 1997. A challenging exploration of the environmental and social issues raised by the drive for utility deregulation. Smeloff is the former chairman of the Sacramento Municipal Utility—which shut down its nuclear reactor and turned instead to solar electricity and conservation.

Solar Energy Research Institute Staff. *A New Prosperity.* Andover, MA: Brick House Publishing, 1981. This was the Carter Administration's detailed blueprint to obtain 20 percent of America's energy from renewable energy by now. Completed in the closing weeks of the Carter Administration, it was promptly scuttled by President Reagan.

Strong, Steven J. *The Solar Electric House: Energy for the Environmentally-Responsive, Energy-Independent Home.* Still River, MA: Sustainability Press (Distributed by Chelsea Green), 1993. The best book on how to power your house with sunlight, by the engineer who has designed more solar-powered houses than anyone else in the country.

Wilson, Alex and Jennifer Thorne, and John Morrill. *Consumer Guide to Home Energy Savings.* Washington, D.C.: American Council for an Energy-Efficient Economy, 1999. Excellent annual guide to energy-efficient products.

World Resources Institute, United Nations Environmental Programme, United Nations Development Programme, and The World Bank. *World Resources.* New York: Oxford University Press. The eight surveys issued in this series have compiled the most important data on global trends affecting energy, climate, human health, and the environment.

Yergin, Daniel. *The Prize: The Epic Quest for Oil, Money and Power.* New York: Simon & Schuster, 1991. This Pulitzer-Prize-winning history of the glory years of the oil industry is meticulously researched and beautifully written. Reading it allows an outsider to better understand the values and attitudes of the industry's leaders today.

Acknowledgments

Former Senator Gaylord Nelson started me down the path to this book thirty years ago by inviting me to be the national coordinator of the first Earth Day. Gaylord's intuitive vote of confidence in a young kid, still in college, profoundly changed the direction of my life.

My other debts are legion. Every sentence was influenced by interactions over the last thirty years with people I admire and by books from which I learned. The following are a few of those whose tutelage most shaped my attitudes and beliefs about energy and climate change. I list them with trepidation because I know I am unintentionally omitting others who will recognize their own insights sprinkled here and there.

John Adams, Donald Aitken, Bruce Anderson, Ken Arrow, Peter Bahouth, John Berger, Scott Bernstein, Deborah Bleviss, Gail Boyer, David Brower, Lester Brown, Clark Bullard, Ralph Cavanagh, Eric Chivian, Mike and Judy Corbett, Angus Duncan, Alan Durning, Paul and Anne Ehrlich, Paul Epstein, Judith Espinoza, Chris Flavin, John Fox, Bob Freling, Charles Gay, Ross Gelbspan, Howard Geller, Al Hammond, Jan Hamrin, Bruce Hannon, Hal Harvey, Paul Hawken, Lisa Hayes, Eric Heitz, Blair Henry, Bob Herendeen, Eric Hirst, John Holdren, Colonel William Holmberg, Henry Kelly, Skip Laitner, Ron Larson, Jonathan Lash, Howard Learner, Paul Leventhal, Maya Lin, Amory and Hunter Lovins, Bill McDonough, Michael McElroy, Bill McKibben, Rose McKinney-James, Gil Masters, Paul Maycock, Alden Meyer, Alan Miller, Fred Morse, Mary Lou Munts, Doug Ogden, Michael Oppenheimer, David Orr, John Perlin, Meg Power, Ken Prewitt, Victor Rabinowitch, Bud Ris, Roby Roberts, Joe Romm, Art Rosenfeld, Rhys Roth, John Ryan, Fred Salvucci, Maxine Savitz, Lee Schipper, Stephen Schneider, Adam Serchuk, Phil Sharp, Rachel Shimshak, Virinder Singh, Bob Solow, Tom Starrs, Steve Strong, Ted Taylor, John Thornton, Sue Tierney, Frank von Hippel, V. John White, Bob Williams, Charles Wyman, and Ken Zweibel.

With an employer less supportive than the Bullitt Foundation, this book and my whole involvement with Earth Day Network would have been impossible. The three major donors to the foundation, Patsy Bullitt Collins, Harriet Bullitt, and Stimson Bullitt, represent the best of philanthropy, combining exuberant creativity with discipline and accountability. The board of the foundation has been a consistent source of sensible advice and personal support. And the staff of the foundation has adapted itself seamlessly to my occasional need to jet to Kyoto or Geneva or Washington, D.C. to put this project together.

The board and staff of Earth Day Network are magnificent. Many of them pulled up stakes and moved to Seattle at a time when we had barely enough money in hand to pay them for a few weeks. From a standing start, they accelerated at warp speed into a major international force. I have too many debts here to list, but the largest is to EDN's executive director, Kelly Evans – one of the finest political organizers of her generation.

My spouse, Gail, lent her unique genius to every stage of this manuscript. Robin Dellabough, my editor at Lark Productions, helped cut through the complexity to find the simplicity on the other side. Dan Sayre of Island Press convinced me to write this book, and then kept things on track with just the right balance of patience and exasperation.

Finally, I want to thank the 1979 -1981 staff at the federal Solar Energy Research Institute for the most intense education of my life. You are the most brilliant, dedicated group I have ever had the privilege of working with. I only wish I'd been able to save you from the barbarians. We could have changed the world.

Index

Acid rain, 5, 30, 110, 126
Actions you can take. *See also* Auto-
    mobiles; Earth Day; Homes;
    Politics; Renewable energy
    ACEEE's checklist for
      homeowners, 89-91
    buying the right car, 51-54
    improving your home's
      efficiency, 79-110
    patronizing non-GCC gas
      stations, 150
    political activity, 6, 48, 117, 127,
      136-137
    recycling, 111-116
    taking control of how you live, 6
    telecommuting, 73-74
    using alternative modes of
      transportation, 68-71
    using renewable energy sources,
      108-110
    voting green, 152-156
Aerators, for faucets, 90, 93, 105
AFUE (annual-fuel-utilization-
    efficiency) rating, 83
Agriculture
    decline in productivity due to
      global warming, 6, 13, 17, 23
    waste products used for fuel,
      45-46
Air conditioners, 86-88
    central air, 87
    choosing the right size, 87
    compared to heat pumps, 88
    EER ratings, 87
    finding a contractor, 87
    replacing filter, 90
    room units, 86-87
    SEER ratios, 86
    use in winter, 94
Alcohol, as biofuel, 44
Alfven, Hannes, 34
Aluminum, recycled, 114, 116
Amazon, fires, 11, 22
American Association for the
    Advancement of Science, 3
American Council for an Energy-
    Efficient Economy (ACEEE),
    54, 85, 89
    checklist for homeowners, 89-91

Consumer Guide to Home
    Energy Savings, Seventh
    Edition, 87, 88, 101, 106
Green Book: The Environmental
    Guide to Cars and Trucks, 54
    Web site, 54, 84, 87, 88, 101, 106
American Forest and Paper
    Association, 149
American Geophysical Society, 10
American Iron and Steel Institute,
    150
American Petroleum Institute,
    118-119
Amtrak, 76-77. *See also*
    Transportation
Anderson, Ray, 116
Animal dung, as fuel, 26
Antarctica
    East Antarctic Ice Sheet, 18-19
    ice shelves, 12, 19
    peninsula, 12, 19
    West Antarctic Ice Sheet, 19-20
Appliances. *See also* Actions
    you can take; Energy; Homes
    buying trends, 7
    clothes dryers, 107-108
    dishwashers, 105-106
    efficiency standards, 80,
      134-135, 140-141
    Energy Star ratings, 80-81, 141
    improvements in efficiency, 7, 78
    lighting, 89, 94-99
    microwaves, 102, 104-105
    refrigerators, 100-102
    stoves and ovens, 102-105
    using energy-saving settings, 89
    washing machines, 106-107
    water heaters, 92-94
    when to replace, 91
Arctic
    drilling for oil, 30
    sea ice, 12
    tundra, 13, 51
Argon gas, 82
Asko dishwashers, 105
Association of International
    Automobile Manufacturers,
    150
Asthma, 5

Atmosphere. *See also* Carbon
 dioxide; Emissions; Global
 warming
 as part of greenhouse
 effect, 10
AT&T, policy on telecommuting, 73
Automobiles. *See also*
 Emissions; Fuel Cells;
 Industry; Politics
 antilock brakes, 55-56
 buying trends, 7, 49-51
 CAFE standards, 65, 131-133, 138
 carpooling, 68
 choosing a car, 51-54
 CO_2 emission calculator, 52
 commuting, 68-69
 composite materials, 56, 60, 67
 driving distances, compared
 with bicycles, 70
 effect on oil industry, 31
 electric cars, 59-61
 fuel efficiency, 50, 54, 131, 162
 fuel-cell vehicles, 26, 31, 43, 44
 *Green Book: The Environmental
 Guide to Cars and Trucks*, 54
 greenhouse gas standards, 141
 hybrid cars, 49, 61-64
 hypercars, 67
 safety, 55-56
 sport utility vehicles (SUVs), 7,
 48, 50-54, 56, 57-58, 71, 132-
 133, 162

Bagasse, 45
Baker, James, 13
Ballard Power Systems, 44, 66. *See
 also* Automobiles; Fuel cells
Bangladesh, monsoon season, 11
Belgium, use of coal, 126
Berry, Wendell, 135
Better Business Bureau, 86
Bicycles, 68, 70-71. *See also*
 Transportation
Biofuels, 45-46, 122, 128. *See also*
 Emissions; Energy;
 Renewable energy sources
 agricultural waste products, 45
 biomass, 6, 26, 45-46
 landfill, 45
 methanol, 45
 paper, 6, 45

reduction of CO_2 emissions, 14
Biomass. *See* Biofuels
Boycott of GCC companies, 147-152
BP (oil company), 67, 150
BTU, 36
 in air conditioners, 86-87
 in walking vs. cycling, 70
Building codes, 140-141
Bush Administration, 121

California
 competition among utilities, 109
 largest car market, 65
 oil fields, 31, 120
 SULEV pollution standards, 62
 use of renewable energy, 102,
 111
 wind farms, 36
Carbon dioxide (CO_2), 10. *See also*
 Carbon fuels; Emissions;
 Global warming
 absorption and emission by
 trees, 13-15, 21-23, 46
 in the air we breathe, 10
 breakdown of emissions (chart),
 15
 from electricity use, 108
 increasing levels, 7, 13, 17
 methods of predicting emissions,
 17
Carbon fuels, 25-32. *See also*
 Coal; Emissions; Natural
 gas; Oil
 burning faster than plants can
 absorb, 13-15, 21-22, 46
 Global Climate Coalition, 147-152
 political support, 117-123
Carbon sink, 21
Carson, Rachel, 16
Carter Administration, 122, 124
Caulking, 81, 90
Center for Resource Solutions, 109
CFC, 4, 14, 107
Charcoal, as biofuel, 45
Chernobyl, 7
Chevron, 118-119, 150, 151
China
 solar-cell incentive programs, 126
 trends in coal use, 30, 126-127
 wind resources, 39
Cholera, 24

Churchill, Winston, 32
Clean Air Act, 118, 126, 140, 146, 152
Clean Energy Agenda, 125, 133,
    162-163
Clean Water Act, 146, 152
Climate. *See also* Emissions; Global
    warming
    as part of Earth Day 2000, 5
    changes in 1998, 10-11
    heat, extreme, 23
    role in spread of infectious
        disease, 23-24
    warming, 16
    weather events, 11-12, 16-17
Clinton Administration, 36, 65, 124,
    125, 132
Clothes dryers, 89, 107-108. *See also*
    Appliances; Homes
    moisture sensor, 108
Coal. *See also* Carbon fuels;
        Emissions; Global warming;
        Industry; Politics
    as a fuel, 28-30
    banning of, 28, 126-127
    "clean" coal, 29
    cost, 110
    emissions, 15, 29, 85, 125
    miners, 127
    mining, 5, 28-29, 126-127
    smoke from, 126
    survival of old coal plants, 140,
        162
    use in generation of electricity,
        110
Colorado, law requiring disclosure of
    electricity sources, 108
Combined heat and power, 140
Compact fluorescent lamps, 89, 95,
    96-99. *See also* Lighting
    as alternative to
        incandescent lamps, 99
    list of brands, 97
    use of mercury, 98
Commuter Choice program, 69
Commuter congestion, 68-69, 72, 73
Conduction of heat, 79, 101. *See also*
    Homes
*Consumer Guide to Home Energy
    Savings*, Seventh Edition, 87,
        88, 101, 105
Consumer Reports, 86, 97, 105

Convection of heat, 81. *See also*
    Appliances; Homes
    in ovens, 103
Conventional ovens, 103. *See also*
    Appliances
Cooper, Senator John Shuman, 152
Copenhagen, bicycle policy, 71. *See
    also* Transportation
Coral reefs, 13
Cornell University, 26
Corporate Average Fuel Efficiency
    (CAFE) standards, 65,
        131-133, 138
Cousteau, Jacques, 23
Curitiba, Brazil, public
        transportation system, 69
Cuyahoga River, 2

DaimlerChrysler, 44
    avoidance of CAFE standards,
        132
    plans for fuel-cell car, 59, 66
    withdrawal from GCC, 150
Daughters of the American
        Revolution, 145
Daylighting, 99
DDT, 14, 145
Decentralized electricity, 27
    use of fuel cells in, 43
Democratic party, 136, 152-153
Denmark, 39, 71
Department of Energy. *See* U.S.
    Department of Energy
Desert Storm War, 50
Diesel engines. *See also* Automobiles
    in cars, 65
    compared with gasoline engines,
        65
    in trains, 75
Dishwashers, 89, 105-106. *See also*
    Appliances; Homes; Water
Drake, Colonel E.L., 16
Dunkirk hot water systems, 84

Earth Day. *See also* Actions you can
        take; Clean Air Act;
        Environmental Protection
        Agency; Politics
    Clean Energy Agenda, 125, 133,
        162-163
    desire for car-fuel efficiency, 50

Earth Day 2000, 5, 113, 156
Earth Day Foundation, 165
Earth Day Network, 164
EPA and Clean Air Act, 118
focus on consumption of fossil
    fuels, 35
founding of, 1, 2-3, 118,
    144-147, 154
Gaylord Nelson, 144
list of sponsors, 156-161
movement, 7, 8, 145
organized labor as major source
    of funding, 145, 154
recycling, 113-115
Edison Electric Institute, 150
Edison, Thomas, 95
Ehrlichman, John, 153
Eklund, Sigvard, 33
Electric cars, 59-61
Electricity. *See also* Appliances;
    Coal; Energy; Fuel cells;
    Hydrogen; Wind power
    consumption from refrigerators,
    100
    electric cars, 59-61
    emissions, 15, 108-109
    fuel cells, 65-67
    green power, 109-110
    Green-e logo, 109-110
    hybrid cars, 61-64
    identifying sources, 108-110
    lack of access to, 26-27
    in lighting, 94-99
    national average electricity rate,
    96
    stoves, 102-103
    water heaters, 92
    from wind, 38-39, 109
Electric stoves, 102-103
Electric Vehicle Association of the
    Americas, 60
Emissions. *See also* Carbon dioxide;
    Carbon fuels; Global warming
    from aircraft, 74-75
    carbon dioxide, 8, 14-15, 44,
    66, 123
    from cars, 50
    from coal use, 15, 29, 85,
    126-127
    from electricity, 108-109
    from forests, 21

fossil fuel industry's prediction
    of 'collapse,'119
from gasoline, 15, 66
greenhouse gases, 10, 30, 48, 66,
    109
from homes, 77
from oil, 15
reduction of, 14, 99, 123-127,
    135-142
scenarios for the year 2100, 29
sulfur, 5, 126
Endangered Species Act, 146
Energy
    as part of Earth Day 2000, 5
    atmospheric, 10
    biofuels, 45-46
    carbon-based fuels, 28-31
    centralized programs, 27
    clean energy checklist, 89-91
    commercialization, 120-121
    decentralized programs, 27
    effect of appliance efficiency
        standards, 135
    effect of recycling, 114
    efficiency output, 7
    how eating less meat saves
        energy, 104
    hydrogen, 41-44
    Internet calculator, 84
    nuclear power, 33-35
    renewable energy, 14, 27, 36-37,
        41, 108-110
    solar energy, 36-38
    use in industry, 139
    use in the home, 92-93, 101, 103,
        105
    wind power, 38-41
Energy audit, 90
Energy Efficiency Ratings (EER), 87
Energy Policy Act, 141
Energy Star program, 80, 87, 89,
    90, 96, 141
Environment. *See also* Climate;
    Emissions; Global Warming;
    Politics
    bipartisan commitment, 152-153
    consequences of global
        warming, 5-6
    movement, 1, 2, 4, 118, 152
    part of who we are, 127
    values, 48, 110

Environmental Defense Fund, 52
Environmental Protection Agency
        (EPA), 2, 80, 96, 118, 140,
        146, 153
Epstein, Dr. Paul, 24
Erdrich, Louise, 138
Erosion, 23, 146
Europe. *See also* Politics
    appliances compared with
        American, 105, 107
    corporate custody of products,
        113
    energy use, 14
    gas taxes increase car efficiency,
        133
    industrial reduction of green-
        house gas emissions, 139
    transportation, 68-70, 75
Evergreen Lease, 116
EV1, 59-61. *See also* Automobiles;
    Electricity
Extinction, of animals and plants, 3,
    5, 13, 46
ExxonMobil. *See also* Automobiles;
    Global Climate Coalition;
    Industry; Oil
    membership and influence in
        GCC, 150, 151
    opposition to solar energy, 118-
        119
    Valdez oil spill, 30

Federal Coastal Zone Management
    Act, 146
Federal Renewable Portfolio
    Standard, 139
Fish catches, 146
Fission energy, 34. *See also* Nuclear
    power
Floods
    flood plains, 16
    weather events, 5, 11, 16
Florida
    coastline, 18
    fires, 11
    Hurricane Andrew, 17
    Hurricane Mitch, 11
Florida Solar Energy Center, 92
Fluorescent lamps, 89, 95, 96-97
    compact, 89, 95, 96-99
    standard, 94-95

tube, 95
tungsten, 95
use of mercury, 98
Ford Motor Company, 57, 59
    plans for fuel-cell vehicle, 66
    withdrawal from GCC, 150
Forests
    Amazon fires, 11, 22
    CO_2 absorption and emission, 13-
        15, 21-23, 46
    fires, 11
    logging, 146
    northern forests (North America,
        Siberia), 22
    Pacific Forest Trust, 22
    timber companies, 21
    tree-planting campaigns, 21, 23
    tropical rainforests, 21-22, 51
Fossil fuels. *See* Carbon fuels
France, use of coal, 126
Fromm, Erich, 36
Fuel cells, 26, 41-43, 122, 127. *See
    also* Automobiles; Hydrogen
    in automobiles, 26, 31, 42, 44, 58,
        65-67
    in combined heat and power
        plants, 140
    methanol, 42
    origin, 43
    phosphoric acid cell, 44
    proton-exchange membrane
        (PEM), 43
    in trains, 75
Furnaces, 78, 83-86. *See also* Homes
    choosing the correct size, 83-84
    efficiency ratings, 83
    finding a contractor, 85
    gas vs. oil, 84
    replacing the filter, 90
    thermostats, 86, 87, 91, 105

Gas. *See* Natural gas
Gasoline, 14, 41, 44. *See also*
        Automobiles; Carbon fuels;
        Oil; Petroleum
    American consumption, 72
    compared with diesel engines, 65
    emissions, 15, 66
    greenhouse gas standards, 141
    leaking from service station
        tanks, 31

taxes, 132, 133, 139
true costs, 132
GE Lighting, 97
General Motors, 57, 59-61, 63-64, 132, 150. *See also* Automobiles; Electric cars; Hybrid cars; Politics
partnership with Toyota, 64, 66
Geophysical Fluid Dynamics Laboratory, 17
Geothermal power plant, 122
Glaciers, 12, 13, 18
alpine and polar, 18
Glacier National Park, Montana, 12
Glass. *See also* Homes; Windows
low-e glass, 82, 84
recycled, 114
Global Climate Coalition (GCC), 147-152, 155
exploitation of scientific debate, 148
list of members, 151-152
"Global Climate Science Data Center," 119
Global warming, 9-24. *See also* Actions you can take; Biofuels; Carbon dioxide; Climate; Coal; Emissions; Energy; Global Climate Coalition; Industry; Politics; Renewable energy sources
as myth, 4, 123, 147
burning of coal and oil, 46, 126-127
compared to tobacco industry, 147
eating lower on the food chain, 104
effect on ozone layer, 24
effect on public health, 23-24
environmental consequences, 5, 10-14
extraction of hydrogen as cause, 42
green power vs. fossil fuels, 110
greenhouse effect, 10, 12, 14, 48
greenhouse gases, 10, 14, 47, 28, 75, 109, 135
Kyoto Treaty, 47, 135-142
role of raw materials, 111

warming trends, 10, 12-13
weather events, 16-17
Gordon, Gil, 73
Gore, Vice President Albert, 124
Green Book: The Environmental Guide to Cars and Trucks, 54
Green-e logo, 109-110
Green energy, 6, 109-110
Greenhouse effect. *See* Emissions; Global warming
Green Lights program, 96
Gross domestic product (GDP), 14, 119
Grove, Sir William, 43

Haldane, J.B.S., 41
Halogen stoves, 102. *See also* Appliances; Homes
Halogen torchiere lamps, 98-99.
compared with Microsun bulb, 99
dangerous heat levels, 98
inefficiency as light source, 98-99
Harvard School of Public Health, 24
HDPE (high-density polyethylene), 115
Heat, extreme, 11, 23-24. *See also* Global warming; Homes; Water heaters; Weather events
in homes, 79-86
from lamps, 94, 98
surface temperature of Earth, 10, 77-78
HeatMirror™, 82
Heat pumps, 88, 90, 92
Heinz, Senator John, 154
Hewlett-Packard, policy on telecommuting, 73
High Mileage Moms, 72
High-speed electric trains, 75. *See also* Transportation
Home cooling, 86-88, 90. *See also* Air conditioners; Homes
HomeEnergySaver web site, 84
Home heating, 79-86, 90. *See also* Homes; Furnaces; Water heaters; Windows
Homes. *See also* Actions you can take; Appliances; Energy
buying trends, 7

Homes (*continued*)
  checklist for better home
    efficiency, 89-91
  cooling techniques, 86-88
  dishwashers, 105-106
  furnaces, 83-86
  lighting, 94-100
  minimizing transfer of heat, 79-83
  size, in relation to family size, 78
Honda Insight. *See* Hybrid cars
Hot water heaters. *See* Water heaters
Hurricanes, 11, 12, 17. *See also*
  Weather events
Hybrid cars, 61-64, 133. *See also*
  Automobiles; Electricity; Fuel
  cells; Hydrogen; Industry
Hydrogen (H2), 41-45. *See also*
  Energy; Fuel cells
  as a fuel, 26, 41, 122, 128
  extraction from fossil fuels, 41, 66
  in fuel cells, 43-44, 66-67, 75
  pumps at filling stations, 66, 67
Hypercars, 67, 122, 128. *See also*
  Automobiles

Incandescent lamps, 89, 95. *See also*
  Lighting
India
  heat strokes (1998), 11
  nuclear power program, 34
  wind turbines, 39
Infectious diseases, 6, 23-24
Insulation in the home. *See also*
  Homes; Windows
  as cooling technique, 86
  in attic, 80, 86, 90
  of walls, 79-81, 91
  of water heater and hot water
    pipes, 89, 93
  of windows, 82
Insurance companies
  reinsurance companies, 17
  weather-related claims, 16
Interface Corporation, 48, 116
International Atomic Energy Agency,
  33, 34
International Energy Agency, 27
International Federation of Red
  Cross Societies, 12
International Panel on Climate
  Change, 29

International Telework Association
  and Council, 73, 74
Israel, use of nuclear power for
  bomb-making capability, 34

Japan. *See also* Kyoto Protocol on
  Global Warming
  coal use, 126
  energy use, 14
  gas taxes increase car efficiency,
  133
Jeffords, Senator Jim, 139, 154
Jenn-Air refrigerator, 101

Kennedy, Robert, 144-145
Krypton gas, 82
Kyoto Climate Conference, 7, 135,
  148-149. *See also* Kyoto
  Protocol on Global Warming
Kyoto Protocol on Global Warming,
  47, 135-142. *See also* Actions
  you can take; Global
  warming; Industry; Politics
  effect of GCC opposition, 148-
  149
  tactics for implementation, 138-
  142, 163-164
  U.S. political opposition, 136-137

Lead, in gasoline, 14
League of Conservation Voters,
  155-159. *See also* Actions you
  can take; Politics; Voting
Lighting, 7, 89, 94-100. *See also*
  Energy; Homes
  compact fluorescent lamps, 89,
  95, 96-97, 98, 99
  generation of heat vs. light, 94
  Green Lights program, 96
  halogen torchieres, 98-99
  incandescent lamps, 89, 95, 96,
  97
  metal halide high-intensity
    discharge lamps, 95, 99
  use of the sun, 99, 100
Louisiana, oil fields, 31, 120
Lovins, Amory, 67
Low-e glass, 82
Low-flow faucet devices, 90. *See also*
  Aerators
Lubchenco, Dr. Jane, 3

Magnetic induction stoves, 103. *See also* Appliances; Homes
Marriott Corporation, 96
McCartney, Paul, 22
Mercury, 29, 98
Mercury vapor lamps, 95
Merrill Lynch, policy on telecommuting, 74
Metal halide high-intensity discharge lamps, 95, 99
Methane, 13, 44. *See also* Emissions; Global warming
    as biofuel, 44
Methanol ("wood alcohol"), 31, 44, 45, 66, 75
Mexico, fires, 11
Microsun lamps, 99
Microwave ovens, 102, 104. *See also* Appliances; Homes
Migration
    of animals, 12, 23
    of people, 27
Minnesota, wind farms, 109
Morris, G.P., 21
Munich Re, 17
Muskie, Ed, 152, 153
Muskie Subcommittee on Air and Water Pollution, 152

Naphtha, 66
NASA, 36, 146
National Appliance Energy Conservation Act of 1987, 135
National Association of Manufacturers, 150
National Center for Atmospheric Research, 11
National Oceanographic and Atmospheric Administration (NOAA), 13, 16
National Renewable Energy Laboratory, 17, 36, 122
National Utility Trust Fund, 139
Natural gas. *See also* Carbon fuels; Emissions; Energy
    as a fuel, 29, 31
    conversion to methanol, 31
    drilling, 31
    emissions, 15
    furnaces, 84, 85
    liquefied natural gas vessels, 31

stoves, 102
    water heaters, 92-94
Natural Resources Defense Council, 134-135
Nelson, Senator Gaylord, 144
Newsprint, recycled, 114
New York Times, The, 119
Nitrogen oxides, 75
Nixon, Richard, 146, 152-153
Nobel Prize, 4
Nuclear power, 33-35. *See also* Energy
    demise of, 33
    fission energy, 34
    half-lives of radioactive materials, 35
    India and Pakistan, 33
    initial IAEA prediction (1974), 33
    output compared with that of solar cells, 128
    plutonium, 5, 33, 35
    potential for bomb creation, 33-34
    reactors, 33
    trends in usage, 34
    waste, 5, 33

Occupational Health and Safety Act, 147
Oil. *See also* Carbon fuels; Emissions; Global warming; Industry; Politics
    as a fuel, 28, 30-31
    birth of industry, 30, 32, 122
    effect on American wealth, 120
    embargo, 7
    emissions, 15
    Exxon Valdez oil spill, 30
    first oil well, 16
    furnaces, 84
    import trends, 51
    oil industry's support of anti-Kyoto Protocol politicians, 138
    political consequences of American dependency, 51
    spillage, 5, 30-31, 145
    uses other than as fuel, 29
Osram Sylvania, 97
Ovens. *See* Appliances; Homes; Stoves and ovens

Ozone. *See also* Carbon dioxide;
    Emissions; Global warming
    CFCs, 4, 14, 107
    from airplanes, 75
    holes, 3

Pacific Forest Trust, 22
Pakistan, nuclear power program, 34
Paper (as energy source), 6, 45, 46
    recycled, 114
Partnership for a New Generation of
    Vehicles (PNGV), 65
Patagonia, 48, 110
Pennsylvania, oil fields, 16, 31, 120
PET (polyethylene terephtalate), 115
Petroleum ("rock oil"), 51, 120,
    122, 149
Philips Lighting, 97
Photoelectrochemistry, 42, 122
Photosynthesis, 14, 38, 46
Photovoltaics (solar cells), 36,
    128, 129, 131
Pimentel, David, 26
Plastic, 115
Plutonium, 5, 33, 35. *See also*
    Nuclear power
Politics. *See also* Actions you can
    take; Industry; Kyoto Protocol
    on Global Warming; Voting
    banning of coal, 126-127
    bipartisanship, 152-154
    defeat of efficiency standards,
        135
    dependence on fossil fuels, 118-
        119
    Desert Storm War, 50
    global warming as myth, 123, 147
    political activism, 6, 48, 117, 135,
        136-137, 143-162
    political goals vs. actual laws
        passed, 125
    presidential agendas, 124-125
    Reagan's dismantling of Carter's
        efforts, 124
    role of government in energy
        revolution, 127, 128, 130-131
Polls, 2, 131
Pollution. *See* Automobiles; Carbon
    dioxide; Carbon fuels;
    Emissions; Global warming
    air, 2

from cars, 50
from electricity, 109
pollution-control technologies, 29
SULEV standards, 62
Polypropylene, 115
Population. *See also* Suburbs;
    Transportation
    existing, 3, 72
    global distribution of wealth, 119
    growth in suburbs and effect on
        commutation trends, 72
Porsche, Ferdinand, 61
Post-consumer recycled content, 115
Princeton University, 17
Publicampaign.org web site, 137

Radiant barriers, 81-82, 86. *See also*
    Windows
Radiant heat, 81
Rainforests. *See* Forests
Raw materials, mining of, 111
Reagan Administration, 36, 38,
    122, 124. *See also* Politics
Recycling, 113-116
    difference between recycling and
        sorting, 115
    ease of recycling as design
        criterion, 113
    post-consumer recycled content,
        115
Reduce-Reuse-Recycle, 111-116
Refrigerators, 89, 100-102, 134.
    See also Appliances; Homes
    correct usage, 101
    Sun Frost, 102
Reinsurance company, 17
Renewable energy sources. *See also*
    Energy; Solar energy; Wind
    power
    as alternative to fossil fuels, 27,
        109-110, 139, 162
    Carter's plan, 124
    defeat in Reagan Administration,
        36-37, 124
    Green-e program, 109-110
    technology, 121
Republican opposition to appliance
    efficiency standards, 79. *See*
    also Politics; Reagan
    Administration
    anti-environment leadership, 153

bipartisanship, 152-154
Resources for the Future, 121
Rio de Janeiro, Earth Summit, 2,
7, 50
Rocky Mountain Institute, 94
Roosevelt, Theodore, 152
Russian grain harvest, 11
R-value, 82

Safe Drinking Water Act, 146
Science, 12
Sea levels, 18-20
Sears, natural gas furnace, 85
Season Energy Efficiency Ratio
(SEER), 86-87
Senate Public Works Committee, 152
Shell Oil Company, 67, 150
Showerheads, high-efficiency, 90, 93
Siberian forests, 22. *See also* Forests
Silent Spring, 147
Silicon, 128
Smog, 31, 65, 110. *See also*
Emissions; Pollution
Sodium lamps, 95. *See also* Lighting
Soil moisture, 17
Solar energy, 127-131. *See also*
Actions you can take; Energy;
Industry; Politics; Renewable
energy sources
as a fuel source, 6, 28, 36, 67,
127-131, 162
as alternative to clothes dryers,
108
bottomless resource, 36-37
cells, 26, 102, 122, 128-131
compared to electronics
industry, 129-130
cost, 128-131
defeat by competing industries
and politics, 36-38, 121-122
extraction of hydrogen from
water (as energy source),
41, 66-67
harvesting, 37
need to phase out coal plants, 126
photovoltaics, 36
potential effect of governmental
investment, 130-131
in reducing CO_2 emissions, 14
in refrigerators, 102
in water heaters, 92, 94

Solar Energy Research Institute. *See*
National Renewable Energy
Laboratory
Solar radiation, 10, 81-82. *See also*
Homes; Solar energy; Sun
Southern Company, 118-119
Spain, use of coal, 126
Sport Utility Vehicles (SUVs), 7, 48,
50-54, 56, 57-58, 71, 132-133,
162. *See also* Automobiles
Stans, Maurice, 145
Steel, recycled, 114
Sting, 22
Stonyfield Yogurt, 48
Storm windows, 81, 82, 91. *See also*
Homes; Insulation; Windows
Stoves and ovens, 102-104. *See also*
Appliances; Homes
conventional vs. convection
ovens, 103
efficient use, 103-104
halogen stoves, 102
magnetic induction stoves, 103
microwave ovens, 102, 104
natural gas vs. electric, 102
self-cleaning ovens, 103, 104
Suburbs. *See also* Automobiles;
Population; Transportation
contribution to commutation
trends, 72
effect on Atlanta, 72
Sun. *See also* Solar energy; Solar
radiation
as lighting source, 99
as a star, 38
contribution to heating of home,
39, 81
creation of wind, 37
role in greenhouse effect, 10
Sun Frost refrigerator, 102
Super Ultra Low Emissions Vehicle
(SULEV), 62. *See also*
Automobiles
Superwindows, 82. *See also*
Windows
Surface temperature of Earth,
77-78
Surface Transportation Policy
Project, 72
SUV. *See* Sport Utility Vehicles
Swidler, Joseph, 94

Tankless water heater, 92
Tax incentives for smart energy, 142
Telecommuting, 73-74
Temperature. *See also* Climate;
    Global warming; Heat
Texas
    fires, 11
    oil fields, 31, 120
    wind turbines, 39
Texas Transportation Institute, 71
Thermostats, 86, 87, 91, 105
Torchieres. *See* Halogen torchiere
    lamps; Lighting
Trains, 74-77
Transportation. *See also*
    Automobiles; Suburbs
    airplanes, 74-77
    bicycles, 70-71
    carpooling, 68
    commutation trends, 71-77
    public, 68
    Surface Transportation Policy
        Project, 72
Trees. *See* Carbon dioxide; Carbon
    fuels; Forests
Trenberth, Kevin, 11
Tropical rainforests, 21-22, 50.
    See also Forests
Tufts University, 48
Tungsten, 95. *See also* Lighting

Udall, Stewart L., 28, 105
U.K. Royal Society, 4
United Kingdom, 126
United States. *See also* Industry;
    Politics
    emissions, 123-125, 135
    energy use, 8, 14
    failure to ratify the Kyoto
        Protocol, 136-138
    reduction of global warming
        pollution, 135
    solar energy leadership, 28
    wealth, 119, 120-123
    wind resources, 38-41
Uranium, 33
U.S. Chamber of Commerce, 150
U.S. Department of Energy, 33, 80,
    85, 135, 141
U.S. National Academy of Sciences,
    4

Vonnegut, Kurt, 146
Voting, 152-156. *See also* Actions you
    can take; Politics

Wall Street Journal, The, 136
Washing machines, 89, 106-107.
    See also Appliances; Homes;
    Water
Waste
    agricultural, 45-46
    from raw materials, 111
    reduction techniques, 112-116
Water. *See also* Water heaters
    coastlines, 5, 18, 20
    in dishwashers, 105-106
    in nature, 5, 16, 146
    in washing machines, 106-107
Water heaters. *See also* Appliances;
    Homes; Insulation
    correct usage, 92-93
    heat pump, 92
    increasing efficiency, 89, 90, 92-
        93
    insulating blanket, 89, 92-93
    pipe insulation, 90, 92-93
    settings, 89, 106
    solar-powered, 92
    tankless, 92
Watt, James, 36
Weather events, 5, 10, 12-13, 16, 17,
    18-20. *See also* Climate;
    Global warming
West Nile-like encephalitis, 24
Whirlpool, 134
Wind farms, 38, 109. *See also*
    Wind power
Windows. *See also* Homes;
    Insulation
    caulking, 81, 91
    HeatMirror™, 82, 84
    insulation, 81-82, 91
    reflectivity, 83
    superwindows, 82
    when to upgrade, 91
Wind power. *See also* Energy;
    Renewable energy sources
    as energy source, 6, 26, 37-39,
        122, 162
    bird deaths, 40
    created by solar energy,
        38-39

effect of Reagan Administration on U.S. wind energy technology, 39
in reducing CO_2 emissions, 14
Wind farms, 38, 109
Wind turbines, 38-39, 122, 128
Winner, Loser or Innocent Victim: Has Renewable Energy Performed as Expected?, 121

Wood. *See also* Forests
as fuel, 6, 26
emissions, 15
heat as radiation, 81
World Bank, 119
World Health Organization, 23
World Petroleum Congress, 149
World Resources Institute, 48

York, natural gas furnace, 85

About Island Press

Island Press is the only non-profit organization in the United States whose principal purpose is to publish books on environmental issues and natural resource management. We are dedicated to providing solutions-oriented information for concerned citizens, public officials, business and community leaders and professionals who are working to solve environmental problems.

In 2000, Island Press celebrates its sixteenth anniversary as the leading provider of timely and practical books that take a multidisciplinary approach to critical environmental concerns and reflect our commitment to make the best information accessible to the environmental community throughout North America and the World.

To find out more about Island Press or to view our on-line catalog, go to *www.islandpress.org*, call 1-800-828-1302, or write to Island Press, Box 7, Covelo, CA 95428.